BN 0792050 4
AF841762
Withdrawn from
University of Brighton
Libraries

Valuing Water, Valuing Livelihoods

Withdrawn from
University of Brighton
Libraries

World Health Organization titles with IWA Publishing

WHO Drinking-water Quality Series

Water Treatment and Pathogen Control: Process Efficiency in Achieving Safe Drinking-water by Mark W LeChevallier, Kwok-Keung Au. (2004)

Safe Piped Water: Managing Microbial Water Quality in Piped Distribution Systems edited by Richard G Ainsworth. (2004)

Protecting Groundwater for Health: Managing the Quality of Drinking-water Sources edited by O Schmoll, G Howard, J Chilton and I Chorus. (2006)

Fluoride in Drinking-water edited by J Fawell, K Bailey, J Chilton, E Dahi, L Fewtrell and Y Magara. (2006)

WHO Emerging Issues in Water & Infectious Disease Series

Heterotrophic Plate Counts and Drinking-water Safety: The Significance of HPCs for Water Quality and Human Health edited by J Bartram, J Cotruvo, M Exner, C Fricker and A Glasmacher. (2003)

Waterborne Zoonoses: Identification, Causes and Control edited by JA Cotruvo, A Dufour, G Rees, J Bartram, R Carr, DO Cliver, GF Craun, R Fayer and VPJ Gannon. (2004)

Pathogenic Mycobacteria in Water: A Guide to Public Health Consequences, Monitoring and Management edited by S Pedley, J Bartram, G Rees, A Dufour and J Cotruvo. (2004)

Water Recreation and Disease: Plausibility of Associated Infections, Acute Effects, Sequelae and Mortality by Kathy Pond. (2005)

Safe Management of Shellfish and Harvest Waters edited by G Rees, K Pond, D Kay, J Bartram and J Santo Domingo. (2009)

Forthcoming: *Animal waste, water quality, human health* edited by A Dufour, J Bartram, R Bos, V Gannon, S Georgiou, D Kay, J Kinzelman, S Kumar, C McPhail, W Robertson and G Toranzos.

For more information please visit www.iwapublishing.com

Valuing Water, Valuing Livelihoods

Guidance on Social Cost-benefit Analysis of Drinking-water Interventions, with special reference to Small Community Water Supplies

Withdrawn from
University of Brighton
Libraries

Edited by
John Cameron
Paul Hunter
Paul Jagals and
Katherine Pond

World Health Organization

IWA Publishing
London • New York

Published on behalf of the **World Health Organization** by

IWA Publishing, Alliance House, 12 Caxton Street, London SW1H 0QS, UK;
Telephone: +44 (0)20 7654 5500, Fax: +44 (0)20 654 5555
Email: publications@iwap.co.uk, Web: www.iwapublishing.com

First published 2011
© World Health Organization 2011

Apart from any fair dealing for the purposes of research or private study, or criticism or review, as permitted under the UK Copyright, Designs and Patents Act (1998), no part of this publication may be reproduced, stored or transmitted in any form or by any means, without the prior permission in writing of the publisher, or, in the case of photographic reproduction, in accordance with the terms of licences issued by the Copyright Licensing Agency in the UK, or in accordance with the terms of licenses issued by the appropriate reproduction rights organization outside the UK. Enquiries concerning reproduction outside the terms stated here should be sent to IWA Publishing at the address printed above.

The publisher makes no representation, express or implied, with regard to the accuracy of the information contained in this book and cannot accept any legal responsibility or liability for errors or omissions that may have been made.

Disclaimer

The information provided and the opinions given in this publication are not necessarily those of the World Health Organization or IWA and should not be acted upon without independent consideration and professional advice. IWA, WHO and the Editors will not accept responsibility for any loss or damage suffered by any person acting or refraining from acting upon any material contained in this publication.

The designations employed and the presentation of the material in this publication do not imply the expression of any opinion whatsoever on the part of the World Health Organization or IWA concerning the legal status of any country, territory, city or area or of its authorities, or concerning the delimitation of its frontiers or boundaries. Dotted lines on maps represent approximate border lines for which there may not yet be full agreement.

The mention of specific companies or of certain manufacturers' products does not imply that they are endorsed or recommended by the World Health Organization or IWA in preference to others of a similar nature that are not mentioned. Errors and omissions excepted, the names of proprietary products are distinguished by initial capital letters. All reasonable precautions have been taken by the World Health Organization to verify the information contained in this publication. However, the published material is being distributed without warranty of any kind, either expressed or implied. The responsibility for the interpretation and use of the material lies with the reader. In no event shall the World Health Organization be liable for damages arising from its use.

The named authors alone are responsible for the views expressed in this publication.

Cover Photographs

Photographs on the cover were all taken in August 2007 in the village of Folovhodwe, Limpopo Province, Republic of South Africa, where the case-study described in chapter 2 was carried out. The African setting in no way implies that the methods and procedures proposed in this book cannot be applied in other continents. All photographs © Robert Bos, WHO.

British Library Cataloguing in Publication Data
A CIP catalogue record for this book is available from the British Library

MIX
Paper from responsible sources
FSC® C007785
www.fsc.org

WHO Library Cataloguing-in-Publication Data
Valuing water, valuing livelihoods / edited by John Cameron ...[et al].

1. Water quality. 2. Potable water. 3. Water pollution. 4. Water supply - standards. 5. Small community water supplies. I. Cameron, John. II. Jagals, Paul. III. Hunter, Paul. IV. Pond, Katherine. V. World Health Organization. VI. International Water Association.

ISBN 978 92 4 156428 1 (NLM classification: WA 675)

IWA Publishing
ISBN: 9781843393108 ISBN: 1843393107
Printed and bound by Bell & Bain Ltd., Glasgow

UNIVERSITY OF
BRIGHTON
07920504
17 NOV 2011
INFORMATION
SERVICES
Withdrawn from
University of Brighton
Libraries

Contents

© 2011 World Health Organization (WHO). *Valuing Water, Valuing Livelihoods*. Edited by John Cameron, Paul Hunter, Paul Jagals and Katherine Pond. Published by IWA Publishing, London, UK.

Foreword

What is it that a farming family in rural Scotland, an Inuit community in northern Canada, a peri-urban slum dweller in South Asia, a young mother in central Malawi, a family at their vacation home in Iceland and a child in an aboriginal family in Australia are likely to have in common? There is a great chance that, to meet their indispensable need for water to drink, to ensure their personal hygiene and to serve their domestic requirements, they rely on a small community water supply.

In fact, a substantial part of the world's population, in high-income, middle-income and low-income countries alike, relies on small community water supplies. While the definition of "small community" will vary by region, what sets these water supplies apart are challenges in ensuring effective administrative, management and technical support structures. Such supplies serve communities that are, by contextual definition, small and frequently remote. They tend to be vulnerable communities, often living in places of climatic hardship, with little access to education and health care and, not uncommonly, at considerable distance from major economic centres. But many peri-urban communities also rely on what can be characterized as small community water supplies.

Living in remote areas may have the benefit of access to more pristine water sources, but poor sanitation may tip the balance with an increased risk of contamination of those sources, and the quality of available groundwater sources cannot always be verified.

Inherent health hazards and their associated risks may be present but will vary from one location to the other. Managing the risks will be a challenge in a setting where facilities are limited, resources constrained and technical know-how comes

© 2011 World Health Organization (WHO). *Valuing Water, Valuing Livelihoods*. Edited by John Cameron, Paul Hunter, Paul Jagals and Katherine Pond. Published by IWA Publishing, London, UK.

at a premium. By definition, small community water supplies cannot benefit from economies of scale. Yet, ensuring access to safe and clean water remains the basic foundation for good health and a key intervention in a primary prevention approach. It can greatly relieve the burden on health services.

Under such conditions, economic evaluation of drinking-water supply options is crucial. It will provide a critical instrument to pave the way for adequate funding streams in support of improvements in access and use. Yet, bearing community vulnerability in mind, a simple analysis of investments required to improve drinking-water supplies with a view to achieving a number of outputs (for example, number of household taps installed) will be insufficient. Small communities derive a host of social benefits from the provision of safe and clean drinking-water, and these have to be valued and made part of the overall equation. Without placing small community water supplies in this livelihood context, its economic case will be hard to make.

This publication, whose production was supported by Health Canada and carried out by a consortium of international experts, gives clear insights into how the principles of social cost-benefit analysis can be turned into practice in the context of small community water supplies.

For small communities, remote or in water-scarce areas, access to safe water is basic to their overall livelihood. Improving their lot in terms of access will require optimal investment in human and material capital. This book is expected to contribute importantly to this goal.

Dr Maria Neira
Director
Department of Public Health and Environment
World Health Organization, Geneva

Acknowledgements

This publication is the output of a collaborative effort undertaken by a Consortium consisting of staff from the University of Surrey, United Kingdom (Katherine Pond and Steve Pedley), the University of East Anglia, United Kingdom (Paul Hunter, John Cameron -currently Institute of Social Studies, Den Haag, The Netherlands- and Chris Edwards) and the University of Johannesburg, South Africa (Paul Jagals currently School of Population Health, University of Queensland, Australia-).

The editors –John Cameron, Paul Hunter, Paul Jagals and Katherine Pond- wish to express their appreciation to all those whose efforts have made the production of this book possible.

The Consortium plus an international group of experts met in Farnham, United Kingdom in February 2007 to discuss the issues and the book structure was developed from this meeting. Subsequently, editorial meetings were held in South Africa in August 2007 and in Geneva in February 2008. The quality of the volume as a whole is due in large part to the review and comments provided by many individuals. Intellectual inputs and review by the following individuals is gratefully acknowledged:

Jay Bhagwan, Director, Water Use and Waste Management, Water Research Commission, Pretoria, South Africa.
Laurence Calame-Haller, Institut F.-A. Forel, Université de Genève, Versoix, Switzerland.
Thomas Clasen, Department of Infectious and Tropical Diseases, London School of Hygiene and Tropical Medicine, London, United Kingdom.
Jack Colford, Associate Professor, Division of Epidemiology, School of Public Health, University of California, Berkeley, Berkeley, CA 94720-7360, USA.

© 2011 World Health Organization (WHO). *Valuing Water, Valuing Livelihoods*. Edited by John Cameron, Paul Hunter, Paul Jagals and Katherine Pond. Published by IWA Publishing, London, UK.

Chris Edwards, Consultant and Senior Fellow, School of Development Studies, University of East Anglia, Norwich, United Kingdom.
Joe Eisenburg, Assistant Professor of Epidemiology, University of Michigan, Ann Arbor, Michigan, USA.
Mutasem El-Fadel, Professor of Environmental Engineering, Director, Environment Division, School of Civil Engineering and Environment, University of Southampton, United Kingdom.
Dima Jamali, Associate Professor of Management, School of Business, American University of Beirut, Beirut, Lebanon.
Eadaoin Joyce, formerly from the Robens Centre for Public and Environmental Health, University of Surrey, Guildford, United Kingdom.
Dawie Mullins, Conningarth Economists, Lynnwoodridge, Pretoria, South Africa.
Steve Pedley, Robens Centre for Public and Environmental Health, University of Surrey, Guildford, United Kingdom.
Luuk Rietveld, Professor Drinking water and Urban water cycle technology Delft University of Technology, The Netherlands.
Helen Riseboro, School of Medicine, Health Policy and Practice, University of East Anglia, Norwich, United Kingdom.
Natasha Robertson, Water and Health Research Unit, University of Johannesburg, South Africa.
Mike Waite, Diauxie, Millgarth, Carr Lane, Sutton on the Forest, York, United Kingdom.
Howard White, International Initiative for Impact Evaluation.

Financial support to the Consortium was provided by Health Canada, First Nations and Inuit Health Branch, Ottawa, Canada. The support of Jennifer Mercer (Health Canada) is gratefully acknowledged. The thorough language editing of the manuscript by Angela Haden (Geneva) is much appreciated. Special thanks are due to Robert Bos (World Health Organization) and Jamie Bartram (currently University of North Carolina) who were instrumental in the successful completion of this project.

Introduction

The aim of this book is to give decision-makers, health professionals and analysts a comprehensive view of the arguments and challenges associated with establishing the value of drinking-water interventions.

The experts who have contributed to this publication provide guidance on assessing the benefits from improving access to safe drinking-water and from reducing the burden of water-related diseases. They show how to compare the value of these benefits to the costs of interventions, with special reference to small-scale drinking-water systems.

The specific focus of this publication is on the socioeconomic appraisal and evaluation of drinking-water interventions. Of course, interventions that combine drinking-water and sanitation improvements will reinforce the benefits from improved drinking-water alone. But while the framework offered here could be applied to sanitation improvements, there are some specific aspects of sanitation that would be better addressed separately.

This book is especially concerned with small-scale drinking-water systems. Such systems are predominantly relevant to rural areas (although the methods described could also be applied, in principle, to large-scale drinking-water systems in urban areas). In any country, communities depending on small systems are the hardest to reach in terms of achieving the water and sanitation targets of the Millennium Development Goals. There is often a difference between the water supplies of urban and rural areas, with rural communities most likely served by a small system. The main differences, however, are in the levels of technology and the institutional arrangements for management, maintenance and protection of water sources. Small drinking-water systems are

© 2011 World Health Organization (WHO). *Valuing Water, Valuing Livelihoods*. Edited by John Cameron, Paul Hunter, Paul Jagals and Katherine Pond. Published by IWA Publishing, London, UK.

also of concern because they are more liable to contamination and breakdown, and therefore pose a permanent health hazard.

To give decision-makers, health professionals and analysts the tools to promote improved access to safe drinking-water, especially for small and vulnerable communities in developing countries, this book discusses this promotion from the point of view of principles and practice, technology and economics, health, livelihoods and ethics.

Chapter 1 explains why it is important to be able to demonstrate the economic value of interventions that will increase access to safe drinking-water, particularly with regard to small-scale interventions.

Chapter 2 shows how, in practice, to carry out an economic assessment of a small-scale drinking-water intervention.

Chapter 3 explores the possibility of low-income communities financing drinking-water interventions. It argues that public resources should be allocated on the basis of an assessment of the full range of social and economic effects of an intervention, rather than just on the basis of narrowly-defined health outcomes.

Chapter 4 outlines the huge problems that small or vulnerable communities throughout the world still face in getting supplies of safe drinking-water. It also emphasizes the benefits of water supply for livelihood activities.

Chapter 5 looks at ways of estimating disease burden within a community and the proportion of disease that may be attributed to a specific environmental risk. In the case of drinking-water, the focus is mainly on diarrhoea.

Chapter 6 explains how to gather livelihoods data to assess the economic changes that result from small-scale drinking-water interventions.

Chapter 7 summarizes the interventions that are currently available to improve communities' access to safe drinking-water through small-scale systems.

Chapter 8 explains how to estimate the financial commitment required to install, maintain and operate a small-scale drinking-water supply system.

Chapter 9 describes how to estimate the physical health impacts of small-scale interventions that give improved access to drinking-water for a target group of people. Clearly, the method could also be applied to other environmental health interventions.

Chapter 10 looks at how cost–effectiveness analysis is done and how it can be used to compare different health interventions.

Chapter 11 discusses the principles of social cost–benefit analysis and shows how they can be applied to drinking-water interventions.

The final Chapter 12 reviews the evidence on drinking-water interventions, available from various studies that use some form of social cost–benefit analysis.

1

Background

Katherine Pond, Stephen Pedley and Chris Edwards

Socioeconomic development is clearly linked to access to safe drinking-water. Recognition of this link is not new. Yet, for the majority of the world's population, lack of access to safe drinking-water continues to be a concern in their daily existence. The *International Decade for Action: Water for Life (2005–2015)* (United Nations General Assembly, 2003) reminds us of the slow progress made over the last century and a half towards what must be the most basic of basic needs in providing a decent quality of life for all globally. As long ago as 1850, the Shattuck report recognized the economic consequences of inadequate water and sanitation:

Withdrawn from University of Brighton Libraries

> "*We believe that the conditions of perfect health, either public or personal, are seldom or never attained, though attainable; that the average length of human life may be very much extended, and its physical power greatly augmented; that in every year, within this commonwealth, thousands of lives are lost which*

© 2011 World Health Organization (WHO). *Valuing Water, Valuing Livelihoods*. Edited by John Cameron, Paul Hunter, Paul Jagals and Katherine Pond. Published by IWA Publishing, London, UK.

> *might have been saved; that tens of thousands of cases of sickness occur, which might have been prevented; that a vast amount of unnecessarily impaired health, and physical disability exists among those not actually confined by sickness; that these preventable evils require an enormous expenditure and loss of money, and impose upon the people unnumbered and immeasurable calamities, pecuniary, social, physical, mental and moral, which might be avoided; that means exist, within our reach, for their mitigation or removal; and that measures for prevention will effect infinitely more, than remedies for the cure of disease." (Shattuck, 1850)*

Since the publication of the Shattuck report, a wealth of evidence has accumulated to show that, where communities lack basic sanitation and use vulnerable and contaminated water, the provision of improved water and sanitation generally leads to a significant reduction in premature mortality and morbidity from water-related infectious disease. But there are other important benefits, sometimes forgotten, that are more difficult to quantify, such as security, privacy and dignity, as well as time saved accessing water. For the purposes of this book, we assume that appropriate technologies exist to achieve these benefits anywhere in the world. Our central concern here is to develop a framework for assessing the socioeconomic value of interventions improving access to safe drinking-water. We focus especially on small systems serving people who would otherwise be difficult to reach. Such people are often missed by large-scale schemes.

Access to safe drinking-water is one of the Millennium Development Goals (MDGs) agreed upon by the world's leaders at the United Nations Millennium Summit in 2000 (see Box 1.1).

The MDG drinking-water target is to halve by 2015 the proportion of people without sustainable access to safe drinking-water. But this is not the first time that the international community has set ambitious targets. In the early 1980s governments enthusiastically embraced the goal of *Water and Sanitation for All by 1990* (United Nations, 1977). At the start of the 1990s, the same goal was restated. In 2004, however, the same number of people lacked access to an improved drinking-water source as in 1990, and these 1.1 billion people included 13 million in developed regions (WHO/UNICEF, 2006).

The 2008 WHO/UNICEF JMP report gave some good news, however. For the first time since reporting began, the estimated number of people without access to improved drinking-water sources had dropped below one billion (WHO/UNICEF, 2008). More than half of the global population now benefits from piped water reaching their homes, and the numbers using unimproved water supplies are going down. The 2010 WHO/UNICEF JMP report showed this trend for drinking-water to be consolidating.

Box 1.1 The Millennium Development Goals

How did they arise? The World Summit for Social Development, held in Copenhagen in 1995, proposed a set of international development targets. These were formally adopted in May 1996 by the Organisation for Economic Cooperation and Development. Subsequently, the United Kingdom Department for International Development also adopted these international development targets, but more modest targets were set by the United States Agency for International Development.

In September 2000, a Millennium Summit held at the United Nations headquarters in New York adopted a set of Millennium Development Goals (MDGs) which were modified versions of the international development targets. In 2001, these MDGs were set out in a road map towards the implementation of the United Nations Millennium Declaration.

What are they? There are eight MDGs with 18 targets or indicators attached to them. All but one of the targets are set for 2015, so we are now well over half-way through the target period. All of these MDGs are aimed at reducing poverty but there are multiple goals because there are multiple dimensions of poverty. Goal 7 aims to "ensure environmental sustainability" and target 10 under this goal aims to "halve by 2015 the proportion of people without sustainable access to safe drinking-water and sanitation." The baseline for the water and sanitation targets is 1990.

Source: Black & White (2004); UNDP (2006).

If progress in achieving access to drinking-water between 2006 and 2015 continues at the same rate as between 1990 and 2008, the global target of halving the proportion without access will be achieved (and actually surpassed) for the developing countries as a whole (Table 1.1). However, two points need to be noted. First, this will still leave some 700 million people *without* access. Second, the rate of improvement needs to be increased for *some* developing regions to achieve *their* 2015 targets. Otherwise these regions (including, most notably, sub-Saharan Africa) will fail to reach the 2015 targets. Over the next decade the population of developing countries is forecast to increase by 830 million, with sub-Saharan Africa accounting for a quarter of the increase and South Asia for another third. Taking into account this population growth, at least an additional 900 million people need access to water by 2015, otherwise these regions will fail to reach the 2015 targets (UNDP, 2006).

Access to safe drinking-water is an essential element of sustainable development, and it is central to the goal of poverty reduction. A recent WHO report (Hutton & Haller, 2004) shows that the total cost of providing safe water varies considerably depending upon the size and location of the target population. In order to achieve the most basic target of halving the proportion of people without sustainable access to an improved water supply by 2015, it has been estimated that developing countries need to spend US$ 42 billion on new coverage (Hutton & Bartram, 2008). The cost of maintaining existing services is estimated to total an additional one billion US dollars for water supply (Hutton & Bartram, 2008).

Table 1.1 The Millennium Development Goals applicable to water (global figures)

	Reference	1990 Actual	2008 Actual	2015 Target
Population with access to an improved water source (%)	WHO/UNICEF (2008)	77	87	89
Population with access to an improved water source (billions)	WHO/UNICEF (2008)	4.1	5.9	6.5
Population without access to safe water (billions)	As implied by the figures in the previous two rows.	1.2	0.9	0.8
Number of people who gained access between 1990 and 2008 (millions)	WHO/UNICEF (2010)		1 774	

A significant challenge to water analysts, including public health engineers, physicians, technicians and economists, is to advise policy-makers on interventions to improve access to safe drinking-water that also produce total benefits greater than total costs. Social cost-benefit analysis, which builds on cost-effectiveness analysis, is a tool to aid this decision-making process. Social cost-benefit analysis is applicable even to small-scale water supplies.

SAFE DRINKING-WATER AS A HUMAN RIGHT

An objection often raised to using economic assessment in decision-making on whether or not to invest in expanding access to safe drinking-water is that a given minimum quantity and quality of drinking-water should be provided as a human right. If this is the case, then surely we do not need to show that drinking-water improvements up to that standard are economically justified by giving a positive rate of return.

As the UNDP *Human Development Report* of 2006 puts it: "ultimately, the case for public action in water and sanitation is rooted in human rights and moral imperatives" (UNDP, 2006, page 42). Article 12 of the International Covenant on Economic, Social and Cultural Rights recognizes "the right of everyone to the enjoyment of the highest attainable standard of physical and mental health" (The Office of the United Nations High Commissioner for Human Rights, 1976). Article 24 of the Convention on the Rights of the Child (Office of the United Nations High Commissioner for Human Rights, 1976) ensures that children are entitled to the enjoyment of the highest attainable standards of health, which requires State Parties to take appropriate measures to combat disease and malnutrition, including within the framework of primary health care (which includes the provision of safe drinking-water).

In 2002, the United Nations Committee on Economic, Social and Cultural Rights, adopted a General Comment on the right to health (United nations Economic and Social Council, 2002). This includes access to safe drinking-water. Regardless of available resources, all States Parties are obliged to ensure that the minimum essential level of rights is achieved, and there is a constant and continuing duty for States to move towards the full realization of a right. This includes ensuring that people have access to enough water to prevent dehydration and disease. The constitutions of more than 90 countries include a reference to the right to water, although such constitutional provision has not been backed by a coherent strategy for extending access to water (UNDP, 2006).

Recently, the United Nations General Assembly adopted Resolution 64/292 confirming that safe and clean drinking-water and sanitation is a human right essential to the full enjoyment of life and all other human rights. Subsequently, the United Nations Human Rights Council affirmed, in its Resolution A/HCR/RES/15/9, that the right to water and sanitation is derived from the right to an adequate standard of living and inextricably related to the right to the highest attainable standard of physical and mental health, as well as the right to life and human dignity (Office of the United Nations High Commissioner for Human Rights, 2010).

When a service or capability is defined as a human right, two problems remain: first, the scope of the human right has to be defined; and second, the human right has to be enforced.

Consider the scope of the right to water. How do we define a minimum standard for water access? Should it be defined in terms of the daily quantity (say, number of litres) to which a household has access? If so, what is that daily amount? What quality standards should this water meet? And what do we mean by acceptable access? Does it mean in the house? Or does it mean within 200 metres from the house? Or within one kilometre from the house?

WHO (2003) defines "no access" as circumstances when it is necessary to travel more than one kilometre or for more than 30 minutes to make a round trip to collect less than 5 litres of water per capita per day. Basic access is considered to be achieved where up to 20 litres per capita per day is available within one kilometre or 30 minutes round trip. Intermediate access is where water is provided on-plot through at least one tap (yard level) and it is possible to collect approximately 50 litres of water per capita per day. Optimal access is a supply of water through multiple taps within the house allowing an average of 100–200 litres per capita per day. Monitoring the supply of water is, however, a problem: "what emerges from research across a large group of countries is that patterns of water use are far more complex and dynamic than the static picture presented in global reporting systems" (UNDP, 2006).

To recapitulate: General Comment 15 on the right to water, adopted in November 2002 by the Committee on Economic, Social and Cultural Rights, sets the criteria for the full enjoyment of the right to water. Yet in 2008 about one in seven of the world's population was denied this basic need. Could the situation be improved by enforcing the human right? The answer is probably no because enforcement of the right to water would not appear to be a feasible option. For example, an attempt in South Africa in 2000 to enforce a right to adequate housing failed, with the Constitutional Court stating that the enforcement of any rights specified in the Constitution depends on the availability of resources. Yet, the right to water and sanitation as now adopted by UN Member States will be a powerful legal instrument to enhance the drive towards the goal of universal coverage, applying the principle of progressive realization.

This means that even if we can agree on a definition of adequacy for access to safe drinking-water, a case needs to be made for expanding sustainable access as compared with competing claims for other poverty reduction measures. That is, the question that will be asked is: does the expansion of access to safe drinking-water have a higher claim on resources than investments in other areas of development? There is, in short, a need for economic assessment of improvements in drinking-water supply.

HOW LACK OF ACCESS TO SAFE DRINKING-WATER AFFECTS WELL-BEING

Unsafe water and sanitation, including lack of hygiene, account for almost one tenth of the global burden of disease (Fewtrell et al., 2007). The use of disability-adjusted life years (DALYs) to measure burden of disease is explained in Chapter 10 of this book. Children under the age of 5 years are particularly susceptible to waterborne disease and suffer the most severe consequences. Other most vulnerable groups include the elderly and pregnant women.

Many life-threatening diarrhoeal diseases are waterborne, so that improving water quality in terms of microbiological contamination is one of the most important contributions of improved water supply to public health. Waterborne and other water-related diseases consist mainly of infectious diarrhoea, typhoid, cholera, salmonellosis, shigellosis, amoebiasis, and other protozoan and viral intestinal infections. Some pathogens causing these diseases are transmitted by water, although other forms of transmission do occur such as person-to-person contact, animal-to-human contact, transmission through food and aerosols, and by contact with fomites (Hunter, 1998). In addition to the dangers posed by pathogenic microorganisms, chemicals such as nitrates, fluoride or arsenic in water can have toxic effects. People who consume water contaminated with these chemicals may not immediately display symptoms of disease, but the long-term effects on their health can be extremely severe, as shown by the example of arsenic poisoning in Bangladesh (Smith, Lingas & Rahman, 2000). In addition, Santaniello-Newton & Hunter (2000) propose a category of diseases that are spread by the daily migration of people to collect water, such as meningococcal disease ("water-carrying disease"). Various non-infectious disorders of the musculoskeletal system resulting from the prolonged carrying of heavy weights, especially during childhood, should also be considered.

A number of studies from low-income countries have indicated that improved access to water – and the resulting increases in the quantity of water or time used for hygiene – are the determining factors of health benefits, rather than improvements in water quality (Curtis & Cairncross, 2003). Providing water security can play a wider role in poverty reduction and improving livelihoods, by reducing uncertainty and releasing resources that can be used to decrease vulnerability. It has been noted that improved domestic water supplies and improved local institutions can enhance food security, strengthen local organizations and build cooperation between people (Soussan, 2003). A water source may be very close to a village but may be of poor quality or only seasonally accessible. In order to reach a source of good quality it may be necessary to travel a considerable distance, thus resulting in less time for other activities (in other words, opportunity costs). In fact, it has been demonstrated that the biggest benefit, in terms of both water and sanitation, is time-saving through better access (Hutton et al., 2007).

In addition to the health benefits and the saving of time and energy, providing safe water can also have an influence on school enrolment and attendance. In many cultures, this particularly affects young school-age girls because, for many poor families, the economic value of a girl's work at home exceeds the perceived returns from schooling. On a wider scale, however, the education of girls is widely attested to lead to a fall in fertility rates and in the next generation's mortality and morbidity rates (World Bank, 2006). Clearly, improvements in water supply increase well-being. But are they a good investment?

This book shows how to assess whether improvements in access to safe drinking-water are a good investment. There are two forms of economic assessment that can be used to do this: cost–effectiveness analysis and social cost–benefit analysis.

WHAT ARE COST–EFFECTIVENESS ANALYSIS AND SOCIAL COST–BENEFIT ANALYSIS?

When WHO identified the need to analyse the costs and benefits of drinking-water interventions as an MDG priority it was clear that there was little work already published on the subject. Earlier work on cost–effectiveness (for example, Walsh & Warren, 1979) had suggested that, of the options for health protection and promotion, water and sanitation interventions were the least cost-effective. This idea persisted for around 20 years until Hutton & Haller (2004) demonstrated, by applying a generalized economic analysis, that water and sanitation interventions are indeed cost–effective. The analysis was applied globally in the *Human Development Report* (UNDP, 2005).

Although the generalized methods were successfully applied at the global level, they do not translate well to the national level. It was clear that there was a need to provide tools on cost–effectiveness analysis and social cost–benefit analysis at a national level to guide policy development. This book describes the methods that can be applied at and below the national level, by people with little or no expertise in economics.

Cost–effectiveness analysis refers to the comparison of the relative expenditure (costs) and physical outcomes (effects) associated with two or more courses of action. In the health sector, cost–effectiveness analysis measures the incremental health outcomes attributable to specific health sector investments, using the direct call on health sector resources as the measure of cost. For WHO, the cost–effectiveness of an intervention is estimated using US$ per case averted, US$ per death averted and US$ per disability-adjusted life year (DALY) saved (Varley et al., 1998). This involves a monetary unit divided by a physical unit. The fact that cost–effectiveness analysis is not measured purely in monetary terms can be seen as an advantage. Generalized cost–effectiveness analysis is used by the Global Programme on Evidence for Health Policy under WHO-CHOICE (**Cho**osing **I**nterventions that are **C**ost-**E**ffective) (http://www.who.int/choice/description/en/). WHO-CHOICE was started in 1998 "with the objective of providing policy-makers with the evidence for deciding on the interventions and programmes which maximise health for the available resources". To achieve its objectives, WHO-CHOICE reports the costs and effects of a wide range of health interventions in the 14 epidemiological subregions, and the results of these cost–effectiveness analyses are assembled in regional databases which policy-makers can adapt to their specific country setting. This has undoubtedly been a useful addition to the tool kits of health policy analysts.

A significant problem with cost–effectiveness analysis is the issue of dealing with wider livelihood benefits. For example, assume that piped water is supplied to a rural village whereas previously the nearest source was 3 km away. In addition to a possible reduction in cases of diarrhoea resulting from the improved access to water, there will be benefits to the households in the form of a saving in time spent in collecting water. It is not straightforward, however, to incorporate livelihood benefits into the WHO generalized cost–effectiveness analysis without assigning values or prices in a common currency to very different benefits. Without such a common currency, only interventions with similar physical outcomes can be compared, virtually ruling out cross-sectoral comparisons.

Social cost–benefit analysis (SCBA) is a framework that allows such comparisons of interventions with complex outcomes. It involves, either explicitly or implicitly, weighing the total expected value of costs against the total expected benefits of one or more actions in order to choose the best or most socially valuable option in terms of value for money. A comprehensive SCBA involves choosing values for all costs and benefits regardless of whether or not they have a market price. In the absence of a clear market price or if the market price is influenced by a powerful public or private agency, then the analyst must choose a price (a shadow price) stating clearly the assumptions that were made in arriving at the value of the shadow price.

To cope with differing patterns of costs and benefits across time, SCBA expresses future costs and benefits of interventions in present-day (year zero) monetary terms. To take account of the value of time ("time is money"), costs and benefits accruing in the future are discounted back to the present by applying a rate of discount to give the "present values" of the costs and benefits (a simple inversion of the calculation used to calculate the value of a present sum of money at any time in the future at a given interest rate). Cost–effectiveness analysis may also use discounting when costs are distributed differently across time. A ranking of interventions can be done by producing ratios of benefits to costs (Hutton & Haller, 2004) or by calculating the net present value of the project by simply subtracting the present value of the costs from the present value of benefits. The ranking can also be achieved by calculating the internal rate of return and this is done by calculating the discount rate which makes the present value of costs the same as the present value of benefits.

One aspect of SCBA that could give rise to controversy is how to value people's time (for example, time saved in collecting water). As discussed by Hutton (2001), assigning a value to people's time could result in a bias towards services for higher income communities. For example, it is common in SCBA to value a life in terms of the future earnings lost. This will mean that, unless a counter-weight is applied to allow for income distribution, the life of a highly-paid person will be valued more than the life of a lowly-paid person of the same age. In its simplest form, SCBA is carried out using only financial costs and financial benefits. A more sophisticated approach to building cost–benefit models is to try to put a financial value on intangible costs and benefits. This involves distributional judgements by the analyst that need to be made explicit and subjected to sensitivity tests, as discussed in this book. Implementation of actions in response to economic assessments is the final step in the procedure.

The main differences between cost–effectiveness analysis and SCBA are summarized in Table 1.2.

Table 1.2 Summary of the differences between cost–effectiveness analysis and social cost–benefit analysis

Cost–effectiveness analysis	**Social cost–benefit analysis**
• helps to select the best possible strategy or technique to follow when the available resources are limited;	• is used to evaluate public expenditure decisions in order to allocate scarce resources in a more efficient way;
• calculates the direct financial cost of reaching specific outcome or output levels and requires one other option for comparison;	• compares all benefits to all costs and can stand alone. (if the benefit/cost ratio exceeds 1, an intervention is socially valuable);
• is typically retrospective;	• is typically prospective;
• gives a micro (community) view of programme activities, outputs or outcomes.	• gives a macro (societal) view.

Improvements to water access – in quantity or quality or both – may create livelihood benefits for all economic activities. Benefits include the funds released for productive investment, and the human time and energy released from water collection or periods of illness. Some of the health and livelihood benefits associated with access to safe drinking-water are discussed in Chapter 6 of this book.

Considering drinking-water as a provider of economic benefits is one way of giving a higher profile to the water MDG target, with a view to attracting development funding. Addressing all the socioeconomic uses of domestic drinking-water and its run-off, and adopting a livelihoods-based approach to drinking-water interventions can provide an economic justification for such interventions relative to others directed at achieving other MDG targets. Where the improvement of drinking-water has been regarded simply as a stand alone matter of health promotion, competing for funding has been confined to the health sector. A more sustainable approach would be to take a broad livelihoods perspective, across all sectors, of the effect of changing drinking-water access and use. Chapters 5 and 6 provide guidance on how to assess the baseline situation as regards the health and livelihood effects associated with water interventions, and the ethical challenges posed in communicating rights to knowledge and intellectual property rights.

A multisectoral economic analysis is more likely to justify cost recovery than the analysis of a single sector. Recognition of economic gains over and above those related to health may mean a greater willingness to pay for improved drinking-water, and a more determined effort to collect fees. This, in turn, may lead to more effective operation, maintenance and repair of the water supply scheme. Cost recovery will be enhanced if improved drinking-water is provided, not only because of its positive effect on health, but also because of its wider economic benefits (Makoni, Manase & Ndamba, 2004).

There are numerous reports of outbreaks associated with small (often rural) water supplies in developed as well as developing countries. Richardson et al. (2007), for example, report on an outbreak of *Campylobacter jejuni* in a South Wales (United Kingdom) rural housing estate which received mains water via a covered holding reservoir. A crack in the wall of the holding reservoir was identified. Contamination with surface water from nearby pasture land was the likely cause of this outbreak. Another problem is that drinking-water can become contaminated following its collection from communal sources such as wells and tap-stands, as well as during its storage in the home. Numerous studies have shown that, taken in isolation, physical improvements to quantity and quality of drinking-water supply have only limited effects on public health, and that household water treatment and safe storage adds considerable value to an integrated approach to improving access to safe drinking-water (Sobsey, 2002).

Generally, the technologies that supply water in small-scale schemes can be technically simple, for example handpump supplies and gravity piped supplies (see Chapter 7). As discussed by Mara (2003), improvements in secure availability of good quality water are required to minimize water-washed transmission of faecal-oral diseases and improve livelihoods. The technologies exist to ensure access to safe drinking-water for all, under local control. In deciding which intervention is most appropriate, values for all costs and benefits associated with the intervention must be estimated.

Justifying funding for small-scale drinking-water interventions is desirable in order not only to reach the Millennium Development Goal targets, but also to achieve the wider development goals of technological and economic sustainability under decentralized, good governance.

The process involved in conducting a socioeconomic evaluation of water interventions in small rural communities consists essentially of five steps:

- Establish a base-line.
- Identify the feasible interventions.
- Estimate the costs of the interventions.
- Estimate the benefits of the interventions.
- Select the best intervention by comparing the social rates of return.

REFERENCES

Black R. and White H. (2004). *Targeting development: critical perspectives on the MDGs.* London, Routledge.

Curtis V. and Cairncross S. (2003). Effect of washing hands with soap on diarrhoea risk in the community: a systematic review. *Lancet Infectious Diseases* 3: 275–281.

Fewtrell L. et al. (2007). Water, sanitation and hygiene: quantifying the health impact at national and local levels in countries with incomplete water supply and sanitation coverage. Geneva, World Health Organization.

Hunter P. R. (1998). *Waterborne diseases, epidemiology and ecology.* Chichester, Wiley & Sons Ltd.

Hutton G. (2001). Economic evaluation and priority setting in water and sanitation interventions. In: Fewtrell, L. Bartram, J. eds. *Water quality: guidelines, standards and health. Assessment of risk and risk management for water-related infectious disease.* London, IWA Publishing, 333–361.

Hutton G. and Haller L. (2004). *Evaluation of the costs and benefits of water and sanitation improvements at the global level.* Geneva, World Health Organization. (For a summary of the report, see http://www.who.int/water_sanitation_health/wsh0404summary/en/, accessed 15 June 2007).

Hutton G. et al. (2007). Global cost–benefit analysis of water supply and sanitation interventions. *Journal of Water and Health* 5: 481–502.

Hutton G. and Bartram J. (2008). Attaining the water and sanitation target. *Bulletin of the World Health Organization*, 86: 19.

Makoni F. S., Manase G. and Ndamba J. (2004). Patterns of domestic water use in rural areas of Zimbabwe, gender roles and realities. *Physics and Chemistry of the Earth, Parts A/B/C*, 29: 1291–1294.

Mara D. D. (2003). Water, sanitation and hygiene for the health of developing nations. *Public Health* 117: 452–456.

Office of the United Nations High Commissioner for Human Rights (1976). International Covenant on Economic, Social and Cultural Rights. Geneva, Office of the United Nations High Commissioner for Human Rights.

Office of the United Nations High Commissioner for Human Rights (1989). Convention on the Rights of the Child. Geneva, Office of the United Nations High Commissioner for Human Rights.

Office of the United Nations High Commissioner for Human Rights (2010). A landmark decision to make the right to water and sanitation legally binding. http://www.ohchr.org/EN/NewsEvents/Pages/RightToWaterAndSanitation.aspx Geneva: Office of the United Nations High Commissioner for Human Rights.

Richardson G. et al. (2007). A community outbreak of *Campylobacter jejuni* infection from a chlorinated public water supply. *Epidemiology and Infection* 135: 1151–1158.

Santaniello-Newton A. and Hunter P. R. (2000). Management of an outbreak of meningococcal meningitis in a Sudanese refugee camp in Northern Uganda. *Epidemiology and Infection*: 124: 75–81.

Shattuck L. (1850). *Report of a general plan for the promotion of public and personal health: devised, prepared and recommended by the commissioners appointed under a resolve of the Legislature of Massachusetts, relating to a sanitary survey of the state*. Boston, Dutton and Wentworth.

Smith A. H., Lingas E. O. and Rahman M. (2000). Contamination of drinking-water by arsenic in Bangladesh: a public health emergency. *Bulletin of the World Health Organization* 78: 1093–1103.

Sobsey M. D. (2002). *Managing water in the home: accelerated health gains from improved water supply*. Geneva, World Health Organization (WHO/SDE/WSH/02.07).

Soussan J. (2003). Poverty, water security and household use of water. *Proceedings of the International Symposium on Water, Poverty and the Productive Uses of Water at the Household Level (Johannesburg)*. The Hague, IRC International Water and Sanitation Centre.

UNDP (2005). *Human Development Report 2005: international cooperation at a crossroads–aid, trade and security in an unequal world*. New York, United Nations Development Programme.

UNDP (2006). *Human Development Report: beyond scarcity–power, poverty and the global water crisis*. New York, United Nations Development Programme.

United Nations (1977). Report of the United Nations Water Conference (Mar del Plata, 14–25 March 1977), United Nations, New York.

United Nations Economic and Social Council (2000). Substantive issues arising in the implementation of the International Covenant on Economic, Social and Cultural Rights. General Comment No. 14 (2000). The right to the highest attainable standard of health. Geneva, United Nations Economic and Social Council.

United Nations General Assembly (2003). *Water for Life Decade 2005–2015*. New York, UN-Water. http://www.un.org/waterforlifedecade/pdf/waterforlifebklt-e.pdf [accessed 4 March 2009].

Varley R. C. et al. (1998). A reassessment of the cost–effectiveness of water and sanitation interventions in programmes for controlling childhood diarrhoea. *Bulletin of the World Health Organization* 76: 617–631.

Walsh J. A. and Warren K. S. (1979). Selective primary health care: an interim strategy for disease control in developing countries. *New England Journal of Medicine* 301: 967–974.

WHO (2005). *Meeting on small community water supply management, Reykjavik, Iceland, 24–26 January 2005.* Geneva, World Health Organization (WHO/SDE/WSH/05.02).

World Bank (2006). *Global monitoring report 2006: Millennium Development Goals–strengthening mutual accountability, aid, trade, and governance.* Washington, World Bank.

WHO/UNICEF (2003). *Report of the first meeting of the Joint Monitoring Programme for Water Supply and Sanitation Advisory Group (Geneva, 23–25 April 2002).* Geneva, World Health Organization.

WHO/UNICEF (2006). *Meeting the MDG drinking water and sanitation target : the urban and rural challenge of the decade.* Report of the WHO/UNICEF Joint Monitoring Programme. Geneva, World Health Organization.

WHO/UNICEF (2008). *Progress on Drinking-water and Sanitation. Special focus on sanitation.* Report of the WHO/UNICEF Joint Monitoring Programme. New York/ Geneva, United Nations Children's Fund/World Health Organization.

WHO/UNICEF (2010). *Progress on Sanitation and Drinking-water, 2010 update.* Report of the WHO/UNICEF Joint Monitoring Programme. New York/Geneva, United Nations Children's Fund/World Health Organization.

2

The practice of economic assessment of small-scale drinking-water interventions

John Cameron and Paul Jagals

This chapter provides a practical and stepwise guide to doing a social cost–benefit analysis. It draws on more detailed work on the economic assessment of small-scale drinking-water interventions. We set out the practical methods for doing such an economic assessment in the following five steps:

- assessing the effect that small-scale drinking-water interventions have on people's livelihoods;
- costing feasible interventions, and assessing their discounted cost–efficiency;
- identifying and measuring the benefits in physical terms, and assessing cost-effectiveness;

© 2011 World Health Organization (WHO). *Valuing Water, Valuing Livelihoods*. Edited by John Cameron, Paul Hunter, Paul Jagals and Katherine Pond. Published by IWA Publishing, London, UK.

- putting values on the benefits and undertaking a social cost–benefit analysis;
- conducting a sensitivity test on a scenario to take account of possible inaccuracies in variables.

These five steps are described below, providing a practical set of tools that can be applied to any small-scale drinking-water intervention in any economy.

To give a sense of application to this process, each step is based on information from a case-study. The case-study chosen is a drinking-water system intervention in the village of Folovhodwe in the north-east of the Limpopo province of South Africa, close to the Zimbabwe and Mozambique borders. The case-study is not offered as typical or representative, rather it offers a range of characteristics that are more challenging than usual for a small-scale drinking-water intervention.

Although the design of the method is robust and the economic assessment could be conducted sitting at a desk, agencies planning or evaluating a small-scale drinking-water intervention should collect primary data beforehand to understand the local context within the target population. The primary data for the case-study were collected using a variety of techniques:

- questionnaire-based surveys;
- direct expert field observation (an important source);
- semi-structured focus groups (which proved a cost-effective technique for collecting the kind of broad parameters required);
- group conversations at standpipe tap points where people were collecting water or washing clothes.

The parameters used here are derived from our own collection of primary and secondary data in the field. We show how such parameters can be used for economic assessment.

ASSESSING THE EFFECT OF DRINKING-WATER INTERVENTIONS ON PEOPLE'S LIVELIHOODS

The first step in any economic assessment of the impact of a drinking-water intervention is to describe the context into which the intervention is introduced. This involves describing the demography of the target group of local people who will be primarily affected by the intervention. This provides a vital factor for scaling economic estimates of variables based on household or individual observations up to aggregate estimates, for example total days of illness prevented. Disaggregation by sex and age is essential for improving the accuracy of such estimates.

We obtained and augmented demographic profiles and maps of the households and the water system. A house-to-house household census was conducted, plus some sample surveys and key informant interviews to gather livelihood data on economic activities and their rates of reward. The data revealed complex patterns of intra-household migration. The Global Positioning System was used to map the water points.

All our survey work was conducted with the full knowledge of local civil society leaders and local government, and no ethical problems were met.

Livelihoods

Data collection on livelihoods in the case-study area was undertaken (in line with the approach recommended in Chapter 6). The aim was to establish the kinds of activities people would undertake with additional time, energy and any other resources released by a small-scale drinking-water intervention. Triangulation of various observations and house-to-house interviews suggested a very low level of monetized economic activity and little produced wealth – the occasional general store and vehicle maintenance or repair workshop were the only signs of commercial activity or investment in technology within the village.

Significant local agricultural activity was observed. For instance, tomatoes are marketed nationally from this area, using natural wealth in land close to the main river. But household surveys suggested very little involvement of the intervention target households in this activity on a continual basis. Similarly, the presence of natural wealth with tourism potential – in the form of a nearby game park – appeared to have little influence on local livelihood activities.

Direct observation and conversations about new housing construction as an indicator of the distribution of produced wealth suggested a heavy influence of remittances from urban areas – often older women were observed living with only their grandchildren or alone in newly constructed sizable houses, some with private water connections. Thus there are both productive and vulnerable people in the population, but they are spatially separated for much of the year. It is difficult, therefore, to talk accurately about the distribution of economic activities and overall labour productivity for many households. Protection against poverty appeared to rest significantly on intra-family remittances and regular State payments of monetary allowances, both for child support and old age pensions.

In terms of human wealth development, there are two primary and one secondary school in the case-study area, and direct observation suggested a high uptake of formal education at both levels. Therefore, the impact on school performance (not necessarily enrolment) is a factor to be considered in the economic assessment.

It was difficult to find evidence of strong social wealth in the area. Support from kin and neighbours, the authority of benign, local chiefs, and the presence of well-attended churches seemingly operated to smooth day-to-day life, reduce vulnerabilities and settle disagreements. But there was a lack of clear collective, deliberative institutions, such as collective meeting places (other than the water taps), or posters or other evidence of advertising events or public meetings.

Some reflection of this limited local social wealth was reflected in the institutional management arrangements for the water scheme. The drinking-water intervention had not been designed or implemented through self-generated local institutions but is a responsibility devolved to the local municipality, though the municipality is not visibly active in running the scheme. Part of the result of this distribution of authority is a widespread sense of powerlessness among local people with respect to undertaking even minor repairs to the system, especially the taps. This causes a general vulnerability of the system to breakdown, and is dealt with in one of the sensitivity tests.

Taken together, these livelihood observations in the case-study area would–at this point–suggest that it would be very unlikely that an economic assessment, based on the local conditions alone, would show a significant net economic benefit or favourable rate of return from a drinking-water intervention. The causal linkages between improved access to safe drinking-water and additional high-value, local economic activities in a rural settlement area with large scale emigration are likely to be weak. If significant economic benefits in terms of high value added do indeed exist in the small-system context, then the economic assessment would need to be extended to a full social cost–benefit analysis, taking account of links to the economy of the whole country (in this case, South Africa) over the long term.

COSTING FEASIBLE INTERVENTIONS AND ASSESSING COST–EFFICIENCY

The framework for the basic costing of a small-scale drinking water system is set out in Chapter 8. It is applied here to assess cost–efficiency.

The first step is to decide on a realistic physical life for the water system; for the case-study intervention, this was set at 20 years (from 1998 to 2017). The significance of a hypothetical moment of closure (in this case in 2017) is that it forces the proponents of all interventions, not just drinking-water interventions, to reflect upon possible environmental impacts that the system might have, such as depleting an aquifer, rather than letting the time-discounting factor erode into insignificance concerns for the more distant future. Environmental impact was not judged to be a significant factor in this scheme because water was drawn from a recharging aquifer at a sustainable rate.

All costs to all affected organizations (public and private), including households, were entered into an EXCEL spreadsheet for the years in which the expenditure actually takes place and the resources are used. The costs were entered for the year when the money was actually spent. A pattern of total costs (including capital, operation and maintenance, and other costs) for the system in the case-study might then look as shown in Table 2.1.

Table 2.1 Synthesized time profile as well as discounted costs for the case-study drinking-water intervention (all costs expressed in terms of prices prevailing in early 2008)

Year	Total costs (thousand rand)	Comments	Discounted value at 3% per annum $Y(0) = Y(t)/(1.03)^t$
1998	1 500	Start of construction	1 500
1999	1 500		1 456
2000	1 500		1 414
2001	1 500		1 373
2002	1 500		1 333
2003	1 500		1 294
2004	175	Taps turned on for normal operation	147
2005	175		142
2006	500	Repairs of teething problems	395
2007	175		134
2008	175		130
2009	175		126
2010	175		123
2011	175		119
2012	500	Replacement of pump	331
2013	175		112
2014	175		109
2015	175		106
2016	175		103
2017	500	Costs to closure less any residual value of remaining assets	285
TOTAL			10 732

The time profile of expenditures for the case-study system was synthesized from the technical specifications of the system, based on standard parameters used by engineers. The pattern shown in Table 2.1 suggests six years dominated by development through construction, and then two years of normal running followed by some minor upgrading as well as maintenance and repairs, while reflecting continual normal running costs, with a hypothetical major maintenance or repair cost in year 2012 for replacement of the pump. The assessment of the intervention ends in 2017 with an endpoint estimate of the costs of restoring non-recharging or slow re-charging aquifers minus the residual value of the remaining assets. These costs were considered both necessary and sufficient to ensure that the system could deliver the planned amount of drinking-water.

In the case-study area, about 7900 households receive water from a system that pumps untreated, but potable, groundwater to a concrete reservoir from where it is gravity-fed to neighbourhood (communal) taps. Capital costs include installing the pump, building the reservoir, assembling and burying piping, and constructing neighbourhood access points (communal taps in this case).

In practice, running the system on a day to day basis is the duty of a villager who is paid 300 rand a month. These costs seem necessary to sustain the system's day to day operational capacity, but are notably insufficient to build the social capital necessary to ensure speedy repairs, local ownership and fair distribution of the water. Running costs to genuinely sustain the system should be considerably higher than this.

It was difficult to get maintenance costs for the case-study. The system seems to be repaired (rather slowly in terms of the taps) rather than receiving preventive maintenance. The pump equipment appears to have functioned well from 2004 to 2007 but, in terms of likely future breakdowns requiring major repairs, the pump is a clear candidate for concern. Therefore provision is made in the costing spreadsheet for the pump being replaced in year 2012. Other than this, maintenance costs have been included at the level considered necessary to sustain the system – higher than actual expenditure in the case-study system because actual expenditure involves some loss of service to significant numbers of people, which fails to meet the political objective of a sustained, high quality supply to all in the target group.

Finally, the system involved no additional expenditure on water transportation or processing by households. Observation showed that households were still using the numbers and types of containers (and occasional wheelbarrows) that they used with the unimproved drinking-water sources.

It is worth noting here that if households paid a tariff or fees for water provision, this would not affect the costing method. In terms of an economic assessment aimed at understanding the social value of an intervention, the concern is with the monetary value of the real resources being used, not who pays the bills.

Discounted cost-efficiency

Discounting is the way economists put a value on time. The discounted value of a cost in the case-study is determined by reducing its value by a discount rate (in the case-study, 3%) for each year between the time when the cost is to be valued (the base year, 1998) and the time that the cost is actually incurred.

To create a level playing field for comparison requires all costs be expressed in terms of one point in time (usually the first year of the intervention, t_0). In the case-study, the heavy expenditure to replace the pump in year 2012 will, at differing interest rates, have present values in 1998 as shown in Table 2.2. All the interest rates are real, in the sense that they ignore price inflation over the life of the system.

Table 2.2 Discounted values in 1998 of 500 000 rand spent or received in 2012 at differing interest rates

Discount or interest rate (%)	**0**	**3**	**5**	**10**	**15**
Present value in 1998	500 000	331 125	252 500	143 600	70 188

The rate of 3% will be used here because it is a rate often used by WHO and other public agencies. It also roughly corresponds to the historic very long term rate of return to low-risk investments. The dramatic power of discounting as a way of putting a value on time is clearly revealed. What prevents discounting from becoming a de facto technical rule of always postponing to tomorrow rather than doing today is the politically set goal of delivering a given level of service to a given group of people as soon as possible – the politics trump the economics in setting a given target in cost–efficiency analysis.

Synthesized costs for the case-study show that the total present value, discounted at 3% per annum give a total cost of 10.7 million rand (Table 2.1). This is the estimated simple cost–efficiency of the system taking account of the time profile of the expenditures as shown in Table 2.1. Any other scheme proposed to provide the target population with safe drinking-water on a sustainable basis should be able to match this total cost in 1998 present value (all expressed in early 2008 prices to remove the element of price inflation).

ASSESSING COST–EFFECTIVENESS

In this section we focus on physical indicators of benefits as measures of cost-effectiveness. Freeing time as a result of the intervention is a general effectiveness indicator, allowing an even wider range of interventions aimed at improving livelihoods and well-being to be compared (see the discussion in Chapters 11 and 12).

In an economic assessment focused on time saving, health benefits come from time freed by fewer episodes of ill-health; that time can now be used for additional livelihood activities. Time may also be made available by preventing premature deaths (discussed separately below). In the simplest case, the number of days that a person is ill in a year is treated as days totally unavailable for any meaningful livelihood activities. But a simple dichotomy of being either totally in or totally out of economic activity ignores the possibility that some activities can continue to be undertaken during an episode of less acute illnesses.

Benefits of reducing morbidity and mortality

The reduction, in the study area, in days affected by illness related to drinking-water was estimated by using days with diarrhoea as a proxy. The estimate was based on studies of the prevalence of the disease 6 months before and 12 months after the intervention. The total number of episodes of diarrhoea prevented was estimated to be 2450 for the 3500 people previously using water from the river. In other words, fieldwork suggested a reduction from 1.1 to 0.3 in diarrhoeal episodes per person per year in the approximately 3500 strong population that had used microbially-contaminated water from the nearby river before the intervention.

The total time savings from diarrhoea reduction can be calculated assuming an estimated average time unavailable for livelihood activities of three days per episode (livelihood activities for our purpose here include any adult activity – both productive and reproductive for the household in economic terms – plus schooling for children). Thus total time available for livelihood activities resulting from the reduction in diarrhoea episodes as a result of the drinking-water intervention for the 3500 people in the area previously using river water can then be calculated as about 7500 (3×2450) days per year (or 20 person–years per year). Those in the study area who had access to a previous smaller scheme are assumed to have no health benefits.

Closely related to time saved in illness, livelihood benefits for those who previously used the river also appear as additional time made available for other activities than caring for a sick person. In the case-study, the time dedicated to caring for sick people was directly linked to the time that the people were ill and was estimated at half a day for every day of illness (that is 3750 days a year or about 10 person–years for the part of the case-study population who previously used the river).

Diarrhoea is also a significant mortality threat for very young children. We considered infants and children under 5 years of age to be at risk if they lived in households that used river water. In the case-study area, about 50 at-risk (i.e. in households relying on the river for their domestic water needs) babies were born in the year before the intervention came into operation, with a further 230 young children being in the highly vulnerable age of 1–5 years. Given the wide access to local primary health care facilities in South Africa, it might be expected that young children to a large extent would be shielded from death resulting from drinking unsafe water. It is also well-known that providing access to safe water is an effective way to prevent early child deaths. Therefore it is assumed for the case-study area that five early deaths are prevented on average per year by the drinking-water intervention.

These five deaths per annum will be added on a cumulative basis to the annual person–years made available in each year over the whole life of the system. No account will be taken that these gains will continue after 20 years, and no account will be taken of the savings in funeral expenses.

Time saved in collecting and processing water

In many circumstances, the largest element in time available for other activities will result from less time spent collecting and (possibly) treating water. Of course, providing better quality water does not necessarily mean decreasing the time and effort involved in collecting water – the better quality water may be further away, depending on the positioning of the taps. But, in general, interventions seek both to improve quality and decrease collection time by providing water from a potable source and creating access points (taps) closer to people's homes (improved access). There might also be time savings in collecting water from the taps, instead of the more remote river, for washing clothes and personal hygiene.

In the case-study, the time saved in collecting water for all activities was very varied, given the large area covered by the water supply system and the wide differences in distances from previous surface water sources. But an average saving of 1.5 hours a day in collecting water per household previously using water from the river seems reasonable. There was no evidence that home-treatment of water was a common practice before the intervention, so no savings (time or produced inputs) were identified. Therefore, total time saved for the previously river-using households in a year was estimated at 330 000 person–hours ($1.5 \times 600 \times 365$). If on average a person spends 10 hours a day on broadly defined, socially valuable livelihood activities (including care for children and the elderly, pre-school learning, formal schooling and community decision-making) that would otherwise have been disrupted by illness, then this is equivalent to 33 000 days or 90 person–years per annum, assuming that the system has operated normally, according to its design specifications.

Savings on health-care expenditure

The estimate of the cost of health sector treatment per episode of diarrhoea is based on the cost of private sector consultation and treatment. In an economic assessment, this can be justified as representing the social cost of treatment by assuming that private sector charges represent market-tested pricing. Consulting a private sector doctor in the broader areas surrounding the case-study area can incur a fee of up to 250 rand. Including medicine, a total cost of up to 1000 rand per treated episode is indicated. Given that the private sector is quite competitive, we treat this as the economic cost to society of health care (in economics terms, the opportunity cost of the resources).

For the population previously using the river, this suggests maximum savings from reducing the number of episodes of diarrhoea by 0.8 episodes per person per year for 3500 people of 2.8 million rand per year if all episodes were treated privately. But in many cases, symptoms would be recognised but medical advice would not be sought or sought only from a nurse in the local primary health care facilities (free to the household but a social cost in public sector resources). Therefore a much lower figure for actual health sector treatment would be reasonable. Assuming this to be the equivalent of about one in seven episodes being treated privately, then the total monetary equivalent of the social cost to households and to the public sector in providing subsidies for public sector health treatment would be 400 000 rand a year.

The complete cost–effectiveness analysis

Now a cost-effectiveness analysis can be undertaken of the impact of the drinking-water intervention in the case-study area. First, discounting will be used for all indicators of effectiveness. For example, preventing an illness now will be considered more socially valuable than preventing the same illness in the future. There is an element of inter-generational bias in favour of the current generation in this approach. But at a discount rate of 3%, this bias should be acceptable, because it is hoped that future generations will have an advantage in terms of access to better medical technology.

We have three different dimensions of effectiveness, measured in three different units:

- reduction in total numbers of episodes of diarrhoea, discounted over the whole life of the intervention;
- greater time available for broadly defined livelihood activities for sick people, those caring for sick people, and time released from collecting and treating water, discounted over the whole life of the intervention;
- monetary or budgetary savings in treatment costs by households and the public sector, discounted over the whole life of the intervention.

A conventional cost–effectiveness approach to the third indicator is to subtract the monetary present value saved in health care from the present value of building, operating and maintaining the system–in other words to treat these savings as a negative cost. This will reduce the total cost of the intervention. The total cost will then be more of a "social" cost, in the sense that the costs taken into account go beyond the direct costs to the agency of building and operating the drinking-water intervention.

Having disposed of this monetized dimension in the costs numerator, the remaining two dimensions are both candidates for the effectiveness denominator. The first is simpler from a health perspective and can be used to compare interventions reducing episodes of diarrhoea. The second includes wider livelihoods data in terms of putting all savings in terms of time saved.

For the case-study, calculations suggested the following values for cost–effectiveness indicators:

- The net present cost is obtained by deducting the present value of financial savings on medical treatment from the present value of capital investment and operation and maintenance costs. At a discount rate of 3% per year, the net present value after this deduction falls significantly to 3.7 million rand (instead of 10.4 million rand, derived earlier in this chapter using a simple cost–efficiency calculation).
- Total discounted reduction in numbers of episodes of diarrhoea was estimated at 25 700. Dividing this figure into the total discounted social costs of 3.67 million rand gives a cost–effectiveness measure of about 150 rand per episode prevented in addition to the costs of health treatment.
- Total discounted gains in terms of time for livelihood activities (released by less illness, less caring for sick people, less time collecting water, and reduced infant mortality) were estimated at 1400 person–years. Dividing

this into 3.67 million rand gives a cost–effectiveness figure of around 2500 rand per person–year of livelihood activity gained.

By themselves, the absolute values of these cost–effectiveness indicators have no meaning. Putting them in a South African context, the sum of money involved in preventing one episode of diarrhoea does not appear cost-effective. The amount of 150 rand is equivalent to almost the weekly wage of a low-paid, full-time employee.

The livelihood time cost–effectiveness indicator looks more cost–effective. A low-paid full-time worker might expect to receive an income of over 12 000 rand a year. So 2500 rand may be an acceptable 'social price' for gaining a whole year of activity. These results are consistent with global economic assessments of small-scale drinking-water schemes, which conclude that a large proportion of the benefits come from time saved in collecting water.

As a final point on using cost–effectiveness analysis to set priorities, there is a need for caution in using cost–effectiveness statistics to make comparisons. Before comparing and making decisions informed by that comparison, it is crucial to ensure that like is being compared with like in terms of the specification of the cost–effectiveness indicator. For instance, there is a need to ask the following questions:

- Have monetary savings been deducted as negative costs in all cases?
- Is the effectiveness indicator identical in specification for all cases?
- Have the same discounting procedures been followed for all variables at an identical discount rate?

Social cost–benefit analysis (an extension of cost–effectiveness analysis) can remove problems of ensuring comparability, not just between drinking-water interventions or across the whole health sector. At its most ambitious, it seeks to compare all interventions coming from every sector that claim to offer improvements in human well-being anywhere in the world. The next section of this chapter is devoted to social cost–benefit analysis.

UNDERTAKING A SOCIAL COST–BENEFIT ANALYSIS

This section demonstrates how to put values on benefits, and use the values to undertake a full social cost–benefit analysis. The cost–effectiveness analysis in the previous section arrived at two estimates of cost–effectiveness: cost per diarrhoea episode prevented; and cost per additional year of human life made available for livelihood activities (including higher quality learning in and out of formal schooling). Social cost–benefit analysis goes beyond this and allows comparisons to be made between all interventions that aim at improving well-being for any group of people on any scale. Clearly, this is important for any agency that wishes to make claims for funding from general development funds beyond that part of the national budget earmarked for the health sector.

Social cost–benefit analysis demands that all costs and benefits be given a monetary equivalent value in terms of prices at a given base year (in our case, 2008 prices). The analyst must choose these values–even where there is no buying and selling in observable markets. Thus the analyst must choose a price that reflects the scarcity of the good or service, for example water in a depleting aquifer. If there is no market but there exists a public sector charge for a good or service, the analyst should reflect on how that charge was decided and how far the charge represents what a competitive market price might be.

Estimating costs and benefits for a full social cost–benefit analysis

Fortunately, social cost–benefit analysis for most small-scale drinking-water interventions is not particularly complex, and robust conclusions can be drawn from the relatively simple framework presented here. For the case-study, costing was provided by an experienced water engineer plus direct observations from the field. Given this, the cost pattern described above is acceptable for the purposes of social cost–benefit analysis.

In terms of the benefits side, we can now treat the savings in health care costs as a monetary benefit, rather than as a saved cost as we did above. We used the price that people pay for private health treatment as a current market-tested monetary value. This is therefore a "shadow" price (in other words, not the real cost paid by most case-study households, but a price representing an open market valuation assuming competition in the private health sector). People, especially in rural areas, predominantly use subsidized public sector clinics or hospitals when they seek treatment, but what they pay does not reflect the full value to South African society of the resources used in diagnosis and treatment.

Using a shadow price has an economic theoretical rationale in social cost–benefit analysis of approximating a market price where forces of demand and supply are freely operating and equated. It also has the advantage of being practical, given that it was impossible to work out a full social costing for the use of local public-sector health facilities to treat diarrhoea without intrusive data collection at local health centres.[1] Even with such data, there would be problems of underestimating full costs, given the way the primary health centres are embedded in a wider, complex public-sector accounting system. This device, of using a chain of equivalent activities (for example different channels for receiving medical treatment) until an open market transaction with a price is identified, is a common practice in social cost–benefit analysis.

A present value of the savings on health treatments was calculated above as about 4 million rand per year. After the intervention, this sum is assumed to be available to support changes in livelihood activities and provide produced assets. These assets can be used to complement additional human time freed by the drinking-water intervention. Thus the freed time could be used more productively in livelihood terms, including possibly purchasing more or better hygiene-related items.

While we now have monetary values for treating an episode of diarrhoea, we have no monetary value for the benefits expressed in terms of gains in person–years of livelihood choices (as a result of time released for livelihood activities through less sickness, less caring for sick people, and time spent collecting water). The starting point is to find answers to the following questions:

- What activities will now be chosen to use the released time?
- Is there a market price for those additional activities?

Given that, in the case-study area, a very low proportion of adults' time is directly sold locally, and much of the time saved concerns people under 18 years of age (who make up over 40% of the population), it might be assumed that there is little monetary value that can be attached to additional time available. So perhaps a monetary equivalent close to zero would be appropriate.

But context is important in developing this aspect of social cost–benefit analysis. First, it is useful in this context to take account analytically of sex and age. In the case-study, assuming that episodes of diarrhoea are evenly distributed by sex and age, then around 25% of time sick will involve adult men, 35% adult women and 40% young people under 18 years of age. For time savings in caring for sick people and collecting water for all its uses, it is

[1] Mullins et al. (2007) make an impressive attempt to calculate the actual total cost of treating an episode of diarrhoea in the public sector in a number of locations in South Africa. Generally, the results show very large differences in cost between treatment in a clinic and a hospital. Our average figure of 1000 rand for a complete treatment lies within the possible range, depending on the proportion of people treated in clinics as outpatients compared with those treated in hospital as inpatients.

assumed that about 75% will be adult women's time, 5% adult men's and 20% that of young people.

In a typical year, adult women will gain a large proportion of the time saved (about 60%) followed by young people under 18 years of age (25%). Therefore, placing a value on time for these two groups is crucial. In the case-study context, given the high level of open unemployment among local males and their limited contribution to work in the home, men over 18 years of age and resident in the case-study area will be given a zero value for their time. The men working as migrants outside the case-study area are vital to the local economy in terms of remittances, but are less likely to suffer from local drinking-water induced illness, care for sick people, or be involved significantly in water collection. Therefore, they do not receive any significant time saving benefits from the drinking-water intervention. Their livelihood activities are therefore assumed to be unaffected by the intervention.

In the case-study area, adult women report that they use time saved to improve the quality of life in the home environment by spending more time in improving hygiene and for better child care. This time has indirect economic value in terms of facilitating the work of other people (including the physiological and psychological impact on rural-urban migrant workers when visiting home) and the schooling of young people.

We will calculate the gains induced by increased studying time when looking at economic gains by young people. The indirect monetary-equivalent gains for supporting other adults in their activities to generate incomes outside the household (in the local economy or as temporary migrants) can be looked at from a wages-for-housework perspective. That is, the additional time freed by the water intervention will enable other household members to be more productive in the wider economy, and this can be expressed in monetary terms. On this basis, it is reasonable to attribute a shadow price of 50 rand a person–day to the additional time made available by the drinking-water intervention (the local wage of a woman working as an employed cleaner). Thus in a typical year, 72 years of adult women's time freed up by the drinking-water intervention will be worth a monetary equivalent of 1.3 million rand ($72 \times 50 \times 365$).

It is difficult to estimate with any precision the qualitative educational gains from the increased total time for studying by people under 18 years of age, resulting from less illness and less time spent caring for people or collecting water, as a result of the drinking-water intervention. However, an order of magnitude for the case-study can be estimated by assuming that:

- there are 200 young people in each one-year cohort who benefit from the drinking-water intervention;

- as a result of the increased study time, energy and adult support attributable to the drinking-water intervention, 10% of each cohort (20 young people) leave formal education having successfully completed one more year than they would have done before the intervention;
- an additional year in formal education is worth, on average, an additional 1000 rand a year over a 30-year working life to each person achieving the extra grade.

Under these assumptions, each young person who achieves an extra year of formal education can expect an increased income valued at a present value in 1998 at 2008 prices of 20 000 rand on a 3% discount rate. Thus 20 young people a year will be credited with a present value equivalent of 400 000 rand to the benefits in every operational year of the intervention following their leaving school.

To put an economic value on infant deaths saved by the intervention will mean that they will be a net cost to their family in terms of consumption costs for much, if not all, of the operational life of the intervention.

Demographically, an additional 65 people (five deaths prevented in each of the 13 years, 2004 – 2017, in which the system is in operation) will be alive at the end of the intervention but who would not have been alive without it. Calculating a value for the net contribution of these 65 people to South African society is a challenge, as the eldest will be only 13 years of age in 2017. Thus significant additional incomes will start around 2020, and from that year more incomes will be added until the oldest start to retire in about 2060, after which total income starts to fall until the last person retires in around 2075. The highest annual total income could be around 1.3 million rand (65 people earning an average of 20 000 rand). Setting this up in spreadsheet format and discounting at 3% per year gives a present value in 1998 of about 15 million rand.

Putting all these benefits into a spreadsheet gives the pattern shown in Table 2.3. Thus the total present value for 20 years of intervention for all types of benefits shown in Table 2.3 in monetary equivalent form is 34 million rand.

Bringing costs and benefits together for analysis

We are now in a position to bring costs and benefits together in a social cost–benefit calculation. Going back to the original cost estimates, the rounded total present value of the costs was 11 million rand. This indicates a net present value (present value of benefits minus present value of costs) of 23 million rand. But net present value in absolute terms is sensitive to scale of operation: generally, a much larger initial investment might be expected to produce a much larger net present value thus confusing comparisons of larger and smaller projects. One way to remove the question of scale is to convert the net present value into a ratio of the present value of benefits (PVB) to the present value of costs (PVC), giving PVB/PVC ratio = 34/11 = 3.1.

Table 2.3 Summary of total discounted benefits

Year	Discounted benefits (million rand)at 3% per year	Comments
1998	15	Discounted flow of income from earnings of saved lives (2020 to 2070)
2004–2016	19	2100 rand each year (400 rand from medical cost savings, 1300 rand from added time for adult women for livelihood choices, 400 rand from income effect of improved school performances). Each year is credited with the same sum of 2100 rand in benefits. The discounting calculation can be simplified as benefits × (sum of all the discounting factors from year 7 to year 19 inclusive). This can be rewritten, as for the case-study, as present value = 2 100 000 $((1/1.03)^7 + (1/1.03)^8 + (1/1.03)^9 + (1/1.03)^{10} + (1/1.03)^{11} + (1/1.03)^{12} + (1/1.03)^{13} + \cdots\cdots + (1/1.03)^{19})$ = 2 300, 00 × 9.2 = 21 160 000
TOTAL	34	

This is a very impressive ratio by any standards and certainly suggests the investment was justified. Generally, a ratio greater than 1.5 is judged to be very satisfactory in assessing public sector investments.

Another way of taking account of scale is to calculate the discount rate that would reduce the net present value to zero. In economics language this is the internal rate of return. Calculating the internal rate of return starts from discarding the assumption of a 3% discount rate and instead calculating the maximum rate of interest that people could afford to pay if a lump sum was borrowed to pay all the costs at the beginning and the whole loan paid back at the end of 20 years. This can be calculated by trial and error using any spreadsheet software. Varying the discount rate and looking at the relative sizes of total costs and benefits (see Table 2.4) will result in the totals of costs and benefits approaching each other; that is, the net present value is getting close to zero and the discount rate is approaching the internal rate of return. At the time of the case-study, the internal rate of return is about 16% per annum – a very creditable rate of return by commercial standards.

Table 2.4 Comparing costs and benefits at varying discount rate

Discount rate	Discounted total costs (in million Rand)	Discounted total benefits (in million Rand)	Comments
15%	7.2	7.7	Need to raise interest rate (internal rate of return) to reduce value of later benefits relative to earlier costs
16%	7.0	6.9	The interest rate (internal rate of return) that almost equates costs and benefits, i.e. the rate the intervention could afford to pay and therefore the higher the better
17%	6.9	6.2	Costs are now higher than benefits so the rate of interest (internal rate of return) needs to fall to increase the value of later benefits relative to earlier costs – that is the intervention can afford to pay a higher rate of interest on a loan

It must be emphasised that this return comes over a period of 20 years. When informing decision-makers, it must always be emphasised that the social cost–benefit analysis estimates are based on estimates of future values of variables, often far into the future, that involve considerable uncertainty.

This may even apply to impact evaluations based on data collected after the end of the intervention if, for instance, they involve estimates of future incomes for people still in school.

This concern with uncertainty about the future (added to doubts about the accuracy in current observations) explains why all the data cited in this chapter are expressed in rounded numbers with two or three significant figures.

Therefore this section must end with a warning. Beware the temptation of offering or demanding spurious accuracy from a social cost–benefit analysis. Citing numbers which give the illusion of much greater accuracy than justified by the procedure for deriving the numbers is very unprofessional and verges on being unethical if it is intended to inhibit discussion of the assumptions being made by the analyst or the likely sampling and measurement errors in the data. Such concerns lead us to the necessity of sensitivity tests, as discussed in the following section.

SENSITIVITY TESTING TO DETERMINE THE ROBUSTNESS OF THE SOCIAL COST–BENEFIT ANALYSIS RESULTS

One of the few truths in economics is that estimates of any mean are accurate only to plus or minus 5% (often attributable to sampling error). Once other forms of inaccuracy are factored in, then the margin of error is likely to be plus or minus 10% or more. Any decision-maker faced with figures rounded to three significant figures and words and phrases such as "about", "estimated", "assumed", "close to", "probably" or "approximately", will be alerted to the fact that they are being offered an imprecise, point estimate of the current situation on the ground. This indicates the need for sensitivity tests.

A sensitivity test constructs a scenario that adjusts some of the values of variables in a social cost–benefit analysis on the grounds that they are comparatively:

- vulnerable to sampling or wider measurement error (in which case both high and low values may be tested to assess impact on cost–benefit ratio or internal rate of return), for example choice of respondents;
- influential on the results of the social cost–benefit analysis because of the sheer scale of their effects (large numbers occurring relatively early in the intervention life), for example delays in construction;
- open to future uncertainty (in the judgement of local key informants or judging by experiences of similar interventions elsewhere), for example breakdown of key equipment;
- of particular concern to decision-makers; that is, some variables have a higher weighting in the political decision than the monetary equivalent value they have been given in the "most likely" social cost–benefit analysis scenario, for example increasing social cohesion;
- of particular concern to people in greater poverty and suffering greater discrimination; that is, some variables have a higher weighting for such people than the monetary equivalent value they have been given in the "most likely" social cost–benefit analysis scenario, for example livelihood damage caused by having to provide "voluntary" labour to construct a new drinking-water system.

Deciding which variables to include in a sensitivity test

The major variables for the social cost–benefit analysis in the case-study are shown in the first column of Table 2.5, with subsequent columns indicating the priority for sensitivity testing. The number of Xs in a cell indicates the sensitivity of the

Table 2.5 Indicative framework identifying variables for sensitivity testing

Parameter variable	Criteria for selection				
	Inaccuracy in measurement	**Scale of Influence**	**Vulnerability to future uncertainty**	**Interest to decision makers**	**Interest to poor people**
Total costs in each year From 1998 to 2017			XXX (histories of poor maintenance locally and globally)		
Livelihood time benefits from fewer diarrhoea episodes			X (if system maintained and population using system remains manageable)		
Livelihood time benefits from caring for fewer sick people	X (once episodes reduction known)		X (if system maintained and population using system remains manageable)		
Proportion of people seeking formal health treatment for diarrhoea episodes			XX (availability and quality of health services)		
Health treatment cost per episode of diarrhoea					
Livelihood time benefits from improved access to water			XXX (rising aspirations to have in-house connections)		

Numbers of infant deaths prevented		XX (through monetary equivalent value attributed)	X (if system maintained and population using system remains manageable)		
Putting a value on each infant death prevented		XXX (distant in time but very high value)	XXX (development of economy)		XXX (source of social security for current generation)
Savings from reduced societal resources needed for health treatment			X (if system maintained and population using system remains manageable)		
Proportion of young people improving school performance	XXX (there is a question of attribution to drinking-water impro vement?)			XXX (especially girls)	
Lifetime income gains from better school performance					X (poorest unlikely to get the highest gains)
Valuation of livelihood time ains differentiating between adult women and adult men					XXX (social justice or inequality dimension)

X = sensitivity of row variable to the factor in the column; XX = very sensitive; XXX = extremely sensitive.

column criteria to changes in the value of the row variable. In economic language, this means the relative degree of elasticity of percentage response of the column variable to a percentage change in the row variable.

Table 2.5 suggests that there are reasons for carrying out sensitivity tests on all of the variables listed. Rather than treat each variable separately, it is often more convenient and more accessible for decision-makers to group the modifications to variables into scenarios with a plausible story to bring out any interrelationships between the variables.

One scenario is presented here that is not dependent on the specific context of the case-study area. It can be applied in almost any situation. Given the positive results of the "most likely" social cost–benefit analysis scenario described above, it seems appropriate to prioritize changes in those benefits most vulnerable to measurement inaccuracy.

The aim of the sensitivity tests is to see whether changes in the variables where accuracy is most in doubt can reverse this positive conclusion. If the social cost–benefit analysis "most likely" result had been negative, it would be logical to see whether modifying the variables in a positive direction within a plausible range would produce a positive result.

In the test scenario, the values of the benefits variables with three XXXs in the appropriate column of Table 2.5 are radically modified to the values shown in Table 2.6.

Putting these modified values into the spreadsheet does not affect the present value of the costs, but it reduces the present value of the benefits to 14 million rand. Therefore, the benefit/cost ratio falls to 1.3 which, while still greater than one, may be less compelling in terms of arguing for a drinking-water project to have priority, as compared with other possible uses of the resources.

Table 2.6 Variables modified to test sensitivity of outcome of social cost–benefit analysis

Parameter variable	Adjustment made
Health treatment cost per episode of diarrhoea	Reduced to 500 rand from 1000 rand
Value of infant deaths prevented	Reduced to zero
Proportion of young people improving school performance as a result of drinking-water intervention	Reduced to 5% of each cohort
Lifetime income gains from better school performance	Reduced to zero

At a conceptual level, this scenario does raise important issues of intergenerational relationships. Any estimates of the state of the world at a time 15 to 45 years in the future must be subject to doubts about the accuracy of the variables involved. The "most likely" scenario puts a considerable value in economic terms on young people's long-term futures and on saving infants' lives. There will always be controversy over putting a value on a human life, and this sensitivity test scenario brings that issue into stark focus.

It encourages decision-makers to take responsibility for long term change and to think about the world of work that will be accessible to the next generation of people.

A LAST WORD

The scenarios offered here are intended to show how social cost–benefit analysis can help the would-be analyst explore issues surrounding a particular small-scale drinking-water intervention in order to offer additional evidence to decision makers. Taken together, the two scenarios show that an economics assessment using social cost–benefit analysis is a tool to assist, rather than dictate, decision-making. Any economic assessment should provoke thought and inform debate, not close the decision-making process.

REFERENCES

Havelaar A. H. and Melse J. M. (2003). *Quantifying public health risk in the WHO guidelines for drinking water quality – a burden of disease approach.* Geneva, World Health Organization (RIVM/734301022/2003).

Mullins D. et al. (2007). *Manual for cost benefit analysis in South Africa with specific reference to water resource development.* Pretoria, Water Research Commission Report No. K8/693.

3

Economic assessments of improvements in drinking-water supply – the global evidence

Chris Edwards and John Cameron

Even if access to safe water is a constitutional human right (as it is in some countries), improvements to drinking-water are likely to be provided only if they can be shown to be a good use of public funds in comparison with the whole cross-sectoral range of possible interventions. This chapter builds on the previous chapter and aims to help policy-makers and other experts understand the global empirical evidence that has been used to criticise and promote drinking-water and closely associated sanitation interventions. It further develops the argument that economic assessments can play a significant role in arriving at an informed judgement of whether or not improvements in water supply are a good use of public funds.

© 2011 World Health Organization (WHO). *Valuing Water, Valuing Livelihoods*. Edited by John Cameron, Paul Hunter, Paul Jagals and Katherine Pond. Published by IWA Publishing, London, UK.

Box 3.1 Terms used for different stages of economic assessment

In this book, we use specific terms to describe economic techniques used at different stages in the project cycle.

- Economic *assessment* is used as an umbrella term when no specific stage in the project cycle is implied.
- Economic *appraisal* refers to economic assessment carried out when possible interventions are being compared, with the objective of prioritizing them for implementation.
- Economic *evaluation* takes place after interventions and usually attempts to capture their total impact, with a view to learning lessons and guiding future priorities for public-sector investment.

Some of the terms introduced in the previous chapter and used throughout this book are described in Box 3.1 and Table 3.1.

There are three forms of economic assessment that can be usefully applied to estimate whether the public sector should finance (completely or in part) improvements in access to safe drinking-water. The simplest is least-cost analysis, which costs proposed interventions that are designed to achieve a specified improvement (a given improvement in quality for a particular population) and finds the least-cost intervention. The second is cost–effectiveness analysis, which is more ambitious in its aims of seeking to compare costs of differing health interventions for different populations against some standard of physical improvement. In this book we use the WHO standard of savings in disability-adjusted life years (DALYs), but any physical measure of improvement (for example, a reduced number of episodes of diarrhoea) would potentially suffice. The third approach is social cost–benefit analysis, which seeks to compare costs across all possible uses of public funds in terms of net benefits to society. Table 3.1 summarizes the three forms of economic assessment.

THE NEED FOR PUBLIC SECTOR INTERVENTIONS

Before looking at the three forms of economic assessment in more detail in Chapters 8, 9, 10 and 11, it is worthwhile answering the following questions:

- Why does the public sector have to be involved?
- Why can the financing of water improvements not be left to households themselves to finance?

Table 3.1 Forms of economic assessment

Form of economic assessment	Characteristics
Least–cost analysis	Estimates the total costs of an intervention, including initial capital investment plus operating and maintenance costs. It may also make risk estimates of the impact of varying the engineering life of the intervention, as sensitivity tests. Costs should be discounted back to their present value to allow comparisons of different forms of intervention to produce a given improvement in water quantity or quality for a given population.
Cost–effectiveness analysis	Takes the costing and divides it by an estimate of a physical benefit to arrive at a cost per unit benefit. The physical benefit may be in terms of a simple directly observable indicator (such as number of episodes of an easily diagnosable illness or symptom prevented). Or it may be a more complex composite indicator such as disability-adjusted life years (DALYs).
Social cost–benefit analysis	May just convert the physical measure in a cost–effectiveness analysis into a monetary value (for example, putting a value on time). But it will usually extend the assessment to include indirect and non-health costs and benefits (for example, monetary value of time saved in collecting water and now used for other purposes). Shadow pricing of costs and benefits where market prices are absent or suspect may be done, and sensitivity tests may be used to assess robustness of estimates of net benefits or internal rates of return. In principle, social cost–benefit analysis allows all forms of drinking-water interventions to be compared with *any* other intervention in any sector that claims to improve human well-being for any scale of population.

A simple but wrong answer to these questions might be that water facilities are public goods and as such should be financed by the public sector in order to ensure sufficient provision.

This answer is analytically flawed because water improvements are not public goods in a strict economic sense. For a rigorous economist, public goods are those goods which, even if consumed by one person, can still be consumed by others. An example of a public good is a lighthouse on a dangerous coast. If the light is shining then "consumption" of that service by one ship does not reduce the consumption available to another ship. This non-rivalry of consumption means that it is impossible to exclude anyone from consumption except at a prohibitive cost. As a result, for the economist, public goods are characterized by non-rivalry and non-excludability in consumption.

By contrast, water services are usually (but not always) private goods, even if not supplied right into the household. If water is provided to a village standpipe, one household's consumption is likely to reduce the amount available to other households – while one container is filling, another cannot be filled. And the less the water available, the more likely there is to be rivalry in consumption and the more likely it is to not be a public good.

Thus, water services are generally public goods, though they may have positive externalities in terms of preventing epidemics of infectious diseases, which may justify an element of subsidy. Similarly, if adding households onto a scheme can be done at low incremental or marginal cost, then again an element of subsidy for all households in the scheme may be economically justified.

But generally water can in principle be bought and consumed exclusively by households, even if an element of public sector subsidy is offered. So the next question arises:

- Why, if water services are not public goods, can households not finance their own facilities?

To answer this, we need to ask two further questions:

- How much do poor rural households spend on water?
- Is this enough to finance improved water supplies?

Whittington & Hanemann (2006) showed that amounts (converted to US$ per month) paid by households to vendors for water in 1998 ranged from 4.4 in Ghana, 6 in Nicaragua, and 7.5 in Pakistan to 13.9 in Côte d'Ivoire. In 2007 prices, this would be equivalent to a range between US$ 6 and US$ 18 per month.

Is this enough to finance improved supplies? As long ago as 1975, Okun, an experienced water supply engineer, thought so. He said that "if daily expenditures made to a water carrier were invested in a proper piped supply, a far more economical and better water service could be provided" (Okun, 1975). One objection to this is that poor people do not, in general, get the whole of their water from vendors. They cannot afford to. There are indications, however, that poor people do spend a significant proportion of their income on water.

The UNDP 2006 *Human Development Report* pointed out that: "The poorest 20% of households in El Salvador, Jamaica and Nicaragua spend on average more than 10% of their household income on water" (UNDP, 2006). The UNDP report was referring to 2004 figures. In that year, the average income per capita of the poorest 20% in these three countries was (according to official statistics) about US$ 430 per year or about US$ 36 per month.[1] This means that, in spending more than 10% of their income on water, poor households spent about US$ 3 to US$ 4 per month on water.

Is this enough to finance water supply improvements? To answer this, we need to know the investment costs of water supply improvements.

Unfortunately, information on the investment costs of water facilities is not available for El Salvador, Jamaica and Nicaragua. For Eastern European and Central Asian countries, the capital cost of protected dug wells serving 100 people is given as about 4000 euros in 2005 (see Environmental Action Programme, 2007, pages 3–8), equivalent to about US$ 5000 or US$ 50 per capita. This compares with an estimate of US$ 48 per capita given by Jamison et al. (2006, Figure 41.1). The match is quite good considering that the estimate given by Jamison et al. is at year 2000 prices, and some allowance needs to be made for price increases between 2000 and 2005.

The annual income of a poor household of six people in El Salvador, Jamaica and Nicaragua is the per capita income of US$ 430 multiplied by six, or about US$ 2580 for the household. Therefore the capital cost of a dug well (at US$ 5000) is equal to almost two years of total household income for the poorest 20% in El Salvador, Jamaica and Nicaragua, and equivalent to almost 20 years of water expenditure (at 10% of total income).

This is likely to be far too much for one poor household to finance, even if the household manages to borrow the money. To illustrate this, assume that a dug well lasts for five years without major maintenance. To repay the cost of US$ 5000 over five years at an interest rate of 5% per annum would mean an annual payment of US$ 1155, whereas at an interest rate of 20% per annum the annual repayment (including interest) would be US$ 1670. As Table 3.2 shows, both these payments are many times the household's annual expenditure on water, which is about US$ 258. And so the dug well is not affordable by one poor household alone.

The dug well can, however, provide water for up to 100 people. The next question is:

- If the 100 people (or 17 households) join together to finance a dug well, does it then become affordable?

---

[1] This is equivalent to US$ 320 per capita. This estimate compares with an estimate of US$ 144 per capita for Latin America given by Jamison et al. (2006, page 772). Thus the estimate of US$ 320 per capita may be on the high side, although again it needs to be noted that the Jamison et al. estimate is at year 2000 prices.

Table 3.2 Can poor households afford improved water supplies? – dug well example

	US$ per capita	US$ per household of six people	Dug well		
Annual income of poorest 20% of households	430	2 580	Capital cost of a dug well serving up to 100 people or about 17 households	US$ 5 000	
Annual expenditure on water at 10% of income	43	258	Repayment factor per year over 5 years	At 5% per year 0.231	At 20% per year 0.334
Repayment costs for a dug well			Repayment costs per year	US$ 1 155	US$ 1 670

The answer is – almost certainly – yes, because even at a high annual interest rate of 20%, the annual repayment cost of US$ 1670 is only about four-tenths of the annual amount being spent on water by the 17 poor households. A small-scale water improvement such as a dug well may, therefore, be affordable – but only if the poor households pool their resources.
This might lead to a further question:

- Is a larger-scale scheme such as a piped water scheme likely to be affordable?

The answer is probably not. According to the Environmental Action Programme (2007, pages 3–13), the cost of a piped water scheme is about US$ 1.6 million.[1]

Even though such a piped water scheme serves up to 5000 people (or 833 households of six people), it is not likely to be affordable unless the poor households spend at least 1.7 times as much as they currently spend on water (see Table 3.3).[2] This 1.7 multiple assumes that people can borrow the money at 5% per annum. If they have to pay a real interest rate of 20% per annum (not unusual in an informal credit market), they would have to spend 2.5 times what

[2] This estimate may be pessimistic, given that a life of only five years is assumed. On the other hand, no allowance has been made for operation and maintenance costs.

Table 3.3 Can poor households afford improved water supplies? – piped water example

	US$ per capita	**US$ per household of six people**	**Piped water**		
Repayment cost divided by household water expenditure:			Capital cost of piped water serving up to 5000 people (or 833 households)	US$ 1,600,000	
For one household	4.5	6.5	Repayment factor per year over 5 years	At 5% per year 0.231	At 20% per year 0.334
For 17 households	0.3	0.4			
Repayment costs for piped water	US$ at 5% per year	US$ at 20% per year	Repayment cost per year	US$ 369,600	US$ 534, 400
Annual repayment cost for piped water	US$ 369, 600	US$ 534, 400			
Repayment cost divided by household water expenditure:					
For one household	1 432.6	2 071.3			
For 833 households	1.7	2.5			

they currently spend. That is, instead of spending 10% of their income on water, they would have to spend 25% of their income on water supplies.

Clearly, on the basis of these figures, there are three problems that poor households will face in financing even small-scale improved water supplies: a loan is likely to be required to finance the improvement; even with a loan, a high element of risk is involved; and a great deal of coordination is required among the households. Therefore, the poorest 20% of households are likely to face problems in financing even small-scale rural water supplies.

In contrast, cost is not necessarily the main barrier to low-cost sanitation improvements. Indeed, Cairncross & Valdmanis (2006) argue against the use of subsidies for such improvements.

But there may well be a strong case for sanitation facilities being provided from public funds on the grounds that these investments are particularly important for environmental quality and health. Without public sector pressure or even financing, external diseconomies are likely to be commonplace, an external diseconomy being the costs imposed by one person (suffering disease from poor sanitation) on another (even though the latter may have adequate sanitation). For an illustration of the external diseconomies from a lack of sanitation, see Box 3.2. In a situation where external diseconomies are common, a private market is likely to provide too little investment (World Bank 1993).

Box 3.2 "External diseconomies" from the lack of sanitation in the United Kingdom in the 19th century

In 1858 the stench of sewage from the River Thames in London forced Parliament to close temporarily. But relatively little was done about sanitation until the mid-1880s. As a result, between 1840 and the mid-1890s, average income in the United Kingdom doubled but child mortality slightly increased. Between the mid-1880s and the mid-1900s per capita investment by the public sector on sanitation increased by more than four times and infant mortality fell, during these two decades, from 160 per 1000 to less than 120.

Source: UNDP (2006).

There still may be a case for public sector support for both water and sanitation improvements, though for slightly different reasons. For water improvements, support from the public sector (at the very least in the form of credit) is likely to be necessary because of the indivisibility of the investment. This is in line with the UNDP position that "in countries with high levels of poverty among unserved [with water] populations, public finance is a requirement for extended access regardless of whether the provider is public or private" (UNDP, 2006). Thus, it is likely that rural water facilities will have to be coordinated – even financed – from outside the households, even though these households may be required (and able) to pay for a major proportion, if not all, of the annual costs.

THE NEED FOR AN ECONOMIC ASSESSMENT

If the public sector is to provide the finance and coordinate the investments for water improvements, then two questions arise:

- Do investments in water interventions give a higher rate of return (in social cost-benefit terms) than other investments, and should they therefore have a greater priority in national and international budgets than they have at the moment?
- In a specific, poor, rural setting, how does a policy-maker decide on the best investment to provide drinking-water? And how can the decision be justified?

It is the job of economic assessment to answer these two questions. As we outlined in previous chapters, there are broadly three methods that are advocated to do an economic assessment of water and sanitation improvements. These are:

- least-cost analysis
- cost–effectiveness analysis
- social cost–benefit analysis.

Least-cost analysis is a method of choosing the appropriate improvement by choosing the one with the lowest cost (see Carlevaro & Gonzales, 2011). However, as Carlevaro & Gonzales admit:

> *when the appropriate [water and sanitation] technologies present differences in the levels or quality of services, a least-cost choice will not necessarily be the one that is economically optimal, as some other appropriate technologies can have benefits that compensate their exceeding costs with respect to the least-cost solution. This is the most common situation, and costing analysis will not provide sufficient information to select the most appropriate technologies. Thus least-cost analysis can be applied when the prioritization decision is solely concerned with choosing between technical interventions offering a similar outcome in terms of improved access to safe drinking-water for the same group of people.*

Cost–effectiveness analysis is widely used by national and international agencies, including WHO. In the health sector, cost–effectiveness analysis is used to select a health intervention which provides a given physical outcome benefit at the lowest cost or the maximum physical outcome benefit for a given budget.

COST-EFFECTIVENESS ANALYSIS AND THE CASES FOR AND AGAINST PRIORITIZING DRINKING-WATER IMPROVEMENTS

A controversy arose as the result of an article by Walsh &Warren (1979). They claimed that prioritization between different uses of health expenditure was an imperative. That meant comparing health interventions, which is a strength of cost–effectiveness analysis. Walsh &Warren claimed that higher health spending was not always associated with better health outcomes, and that health budgets could be spent more cost-effectively. Few disagreed with this view, which was to be endorsed in the World Bank's 1993 *World Development Report* (see Box 3.3).

Box 3.3 Higher health expenditure does not mean better health

At any level of income and education, higher health spending might be expected to yield better health, but this is not the case. The World Bank's 1993 *World development report* showed that there was no relationship between health spending as a percentage of gross national product and health (as defined by life expectancy), after allowing for levels of income and education. The World Bank pointed out that "China... spends a full percentage point less of its GNP on health than other countries at the same stage of development but obtains nearly ten years of additional life expectancy" and that; "Singapore spends about 4 per cent less of its income on health than others at the same level of development but achieves the same life expectancy". By contrast; "... it is possible both to spend more than predicted on health care and still achieve unexpectedly poor results. The United States is an extreme case spending 5 per cent more of GNP than predicted to achieve several years less of life expectancy than would be typical for its high income and educational level".

Source: World Bank (1993).

What *was* controversial about the Walsh & Warren (1979) paper was the case they made for a "selective primary health care" programme and the way in which that case was made. A year earlier, in 1978, a worldwide primary health care movement had been launched under the slogan "Health for All by the Year 2000" at a conference held by WHO and UNICEF at Alma Ata in the, then, USSR (WHO, 1978). In 1979, the Walsh & Warren paper was presented at a Rockefeller Foundation Conference in Bellagio, Italy (reported in Warren, 1988). In it, the authors advocated a selective primary health care programme, and the paper was published in the *New England Journal of Medicine.*

Walsh & Warren argued that infant and child mortality could be reduced most effectively by primary health care that was selective in terms of types of interventions. They claimed that the deaths from many of the most prevalent diseases could be best prevented by immunization, oral rehydration, universal breast-feeding and by antimalarial drugs for African children. In the following years, immunization programmes were adopted. The influence of this approach was reflected in the 1993 *World Development Report* in which the World Bank endorsed an expanded programme on immunization (EPI), stating that the programme could be enlarged still further to include supplements such as vitamin A and iodine, and other vaccines, particularly those for hepatitis B and yellow fever. The World Bank stated that: "in most developing countries, such an 'EPI Plus' cluster of interventions in the first year of life would have the highest cost-effectiveness of any health measure available in the world today" (World Bank, 1993).

The Walsh & Warren approach was widely supported by UNICEF and by a number of bilateral and multilateral donor agencies, but it was also heavily criticized (Warren, 1988). Given the widespread adoption of selective primary health care, why was the approach so heavily attacked?

Walsh & Warren (1979) drew up priorities on the basis of cost–effectiveness calculations. In the paper, health-promoting interventions were ranked in terms of their cost-effectiveness in achieving very specific, physical health outcomes, notably in terms of infant and child mortality. The paper was attacked on two grounds. John Briscoe of the School of Public Health in North Carolina was an important exponent of the group that criticized the paper.

First he argued that such specific physical indicators understated the general health benefits of water and sanitation programmes.[3] He pointed out that a review of the health effects of water supply and sanitation programmes carried out in 1983 revealed that the reduction in the incidence of diarrhoeal diseases in the population at large was typically between 30% and 40%, many times greater

[3] There were other, broader criticisms focusing on issues of organization and power. Smith & Bryant (1988) suggested that the attention "given to the delivery of 'selective' packages of interventions has often diverted energy and resources from the essential task of developing comprehensive, efficient and effective health services"

than the 5% assumed for standpipe water in the Walsh & Warren paper (Briscoe, 1984). This empirical claim suggested that health benefits were spread widely in the population benefiting from a drinking-water intervention, as well as more infant and child deaths being prevented.

The second and more important criticism was that the large non-health benefits generated by water and sanitation improvements were ignored in the Walsh & Warren cost–effectiveness approach. If these non-health benefits (notably time savings in collecting water) were deducted from the costs, the net economic cost of water supply improvements would be much smaller than the gross cost and the picture would be very different (Briscoe, 1984).

Briscoe (1984) argued that, as a result "it is apparent that the cost–effective calculations of the [selective primary health care] approach are fundamentally flawed when dealing with community water supplies" and he complained that "the [selective primary health care] approach in general and the downgrading of water and sanitation, in particular, seem to have been accepted implicitly by many development agencies". He also pointed out that

> *"just three years after the proclamation of the International Drinking Water Supply and Sanitation Decade by the United Nations General Assembly, the Decade is being pronounced 'dead' in some quarters". (Briscoe, 1984)*

This example shows the limitations of cost–effectiveness analysis for deciding between interventions in terms of conceptualizing indicators of benefits (e.g. mortality, incidence of diarrhoeal episodes, time saved in collection), specifying data to be empirically collected and identifying disagreements on the estimates of indicators made from the collected data.

This book advocates both for extending the conceptual range used in cost–effectiveness analysis and for improving the quality of data being collected, especially for small-scale drinking-water interventions.

SOCIAL COST–BENEFIT ANALYSIS AND THE UNDP 2006 *HUMAN DEVELOPMENT REPORT*

The points made by Briscoe (1984) were dramatically endorsed more than 20 years later in studies carried out by Hutton, Haller and Bartram (Hutton & Haller, 2004; Hutton, Haller & Bartram, 2006). These studies placed monetary values on costs and benefits, and claimed high ratios of benefits to costs for water and sanitation investments. These high benefit–cost ratios were highlighted in the 2006 UNDP *Human Development Report*. The Hutton, Haller and Bartram studies seem to support the Briscoe claim that the non-health benefits of water and sanitation improvements are very large indeed when converted into their monetary equivalents. Indeed, according to Hutton, Haller & Bartram (2006), the non-health benefits (especially time savings) formed the vast majority of the total, as shown in Table 3.4.

Table 3.4 Benefits and costs for sub-Saharan Africa from meeting the year 2015 MDG targets for water and sanitation over the period 2005–2015

	Water	**Sanitation**	**Tables in Hutton, Haller & Bartram (2006)**
Number of people getting improved water or sanitation (million)	207	315	Table 14
Annual costs (US$ billion)	0.48	2.19	Table 14
Annual benefits (US$ billion)	0.12	0.31	Tables 17, 18
–health system and patient costs saved	0.11	0.45	Table 19
–value of time saved from less illness	0.27	0.72	Table 20
–value of access time saved	0.84	12.88	Table 21
Total benefits (US$ billion)	1.34	14.36	Table 13
Benefit/cost ratio	2.8	6.6	Table 11
Percentage of total benefits from access time saved	63	90	

It is clear that most of the benefits from water and sanitation improvements come not from improvements in health (for example, from a reduction in illness or death), but from a saving in time in accessing water sources and sanitation facilities. Table 3.4 shows that for water supply improvements in sub-Saharan Africa, 63% of the annual benefits come from time savings and only 37% from savings associated with a reduction in illness.

Whereas the study by Walsh & Warren (1979) had provided decisions-makers with reasons *not* to go ahead with drinking-water improvements, a quarter of a century later the UNDP 2006 *Human Development Report* was pointing to the high benefit–cost ratios to be obtained from such investments. Thus the UNDP *Human Development Report* gives good reasons to go ahead with water improvements prior to many interventions in other sectors of the economy.

As a basis for priority-setting, social cost–benefit analysis is able to compare investments across sectors because all benefits and costs are converted into monetary equivalents.

As Hutton, Haller & Bartram (2006) put it: "these results give to water and sanitation advocates a powerful basis for arguing for increased water and sanitation investments".

The UNDP 2006 *Human Development Report* estimated that the investment outlay needed to reach the MDG targets for water and sanitation with low-cost sustainable technology would amount to about US$ 10 billion a year, whereas the monetary equivalent of the benefits would be well over this, at about US$ 38 billion a year (UNDP, 2006). When the costs are spread over their economic life, the economic return is high. Each dollar spent yields a return of about US$ 8 in costs averted and productivity gained (UNDP, 2006). According to UNDP, these figures probably understate the gains from water investments since they do not capture the benefits from education, from empowering women, from human dignity, or from the reduced anguish and suffering associated with lower rates of child deaths.

The *Human Development Report* states that: "Ultimately, the case for public action in water and sanitation is rooted in human rights and moral imperatives. At the same time, cost–benefit analysis suggests that economic common sense makes a powerful supporting case" (UNDP, 2006).

The fact that water investments give a mix of benefits is clearly something of a political disadvantage. As Walsh (1984, page 1167) said:

> *"A health planner, faced with the charge of improving health with the few resources available, may decide not to make capital investment in water supply and sanitation a top priority.... Possibly it would be more appropriate for the agricultural, or public works, or planning and development department, with collaboration from the health sector, to invest in an improved water supply and sanitation because all these sectors will benefit".*

Water supply improvements provide a mix of health and other benefits. Advocates for water and sanitation improvements are therefore at a disadvantage, because the ministry likely to bear the costs may feel it receives insufficient credit for benefits that are perceived to come under the mandate of a number of other ministries.

THE WAY FORWARD

Few social cost–benefit analyses seem to have been made of improvements to drinking-water facilities, perhaps as a result of the predominance of public health experts in debates. If such analyses exist, they have not made their way into the public domain.

Cost-effectiveness analyses are more widely used by national and international agencies, including WHO. In the health sector, cost–effectiveness analysis is used to select a health intervention which provides a unit of physical output at the lowest unit cost. Thus a physical rather than monetary indicator of output is chosen, and the option which has the lowest cost per unit of output is preferred. Cost–effectiveness analysis can play an important role in comparing different health interventions. The basic data on benefits can often easily be derived from standard health statistics, and calculations and interpretations can be made by non-economists. Cost–effectiveness analysis has consequently been widely applied in analysing different drinking-water interventions. One of the most comprehensive of the cost–effectiveness studies is that of Clasen et al. (2007), and Chapter 10 includes a discussion of that study.

Encouraging decision-makers to rely not only on cost–effectiveness analysis, but also on the more comprehensive information provided by social cost–benefit analysis, is a major aim of this book. Social cost–benefit analysis has the merit of being able to break out of the health sector and offer comparisons with any intervention claiming to improve human well-being. This is important in making claims for better funding of drinking-water interventions from the general public purse – whether they be to improve lives of smaller groups of currently underprovided people in richer economies or to finance general improvements to achieve the health MDG (and assist in the achievement of other MDGs) in poorer economies. The drawback of social cost–benefit analysis is that it is based on highly technical economics concepts, and hence requires the greater involvement of economists – a profession generally not admired for its lucidity and communication skills.

REFERENCES

Briscoe J. (1984). Water supply and health in developing countries: selective primary health care revisited. *American Journal of Public Health*, 74: 1009–1013.

Cairncross S. and Valdmanis V. (2006). Water Supply, Sanitation and Hygiene Promotion. Ch.41 in Jamison et al. (eds.) *Disease Control Priorities in Developing Countries*. Washington, DC, World Bank. Available at: www.dcp2.org/pubs/DCP

Carlevaro F. and Gonzalez C. (2011). *Costing improved water supply systems for low-income communities: a practical manual*. CDROM. Geneva, University of Geneva and World Health Organization.

Clasen T. et al. (2007). Cost–effectiveness of water quality interventions for preventing diarrhoeal disease in developing countries. *Journal of Water Health*, 5: 599–608.

Environmental Action Programme (2007). *Rural water supply and sanitation: technology and cost functions*. Environmental Action Programme Task Force, EU Water

Initiative (http://www.oecd.org/dataoecd/25/2/38080342.pdf, accessed 23 March 2009.

Hutton G. and Haller L. (2004). *Evaluation of the costs and benefits of water and sanitation improvements at the global level.* Geneva, World Health Organization (WHO/SDE/WSH/04.04).

Hutton G., Haller L. and Bartram J. (2006). *Economic and health effects of increasing coverage of low cost water and sanitation interventions.* Occasional Paper for the 2006 Human Development Report.

Jamison et al. eds. (2006). *Disease control priorities in developing countries.* Washington, DC, World Bank.

Okun D. (1975). Review of drawers of water. Economic Development and Cultural Change, 23: 580–583.

Smith D. L. and Bryant J. H. (1988). Building the infrastructure for primary health care; an overview of vertical and integrated approaches. *Social Science and Medicine 9*: 909–917.

UNDP (2006). *Human Development Report: beyond scarcity–power, poverty and the global water crisis.* New York, United Nations Development Programme.

Walsh J. A. and Warren K. S. (1979). Selective primary health care: an interim strategy for disease control in developing countries. *New England Journal of Medicine*, 301: 967–974.

Walsh J. A. (1984). Comment on water supply and health in developing countries. *American Journal of Public Health*, 74: 1167.

Warren K. (1988). The evolution of selective primary health care. *Social Science and Medicine*, 26: 891–898.

Whittington D. and Hanemann W. M. (2006). *The economic costs and benefits of investments in municipal water and sanitation infrastructure: a global perspective.* Berkeley, Department of Agricultural and Resource Economics, University of California, Working Paper 1027). (http://repositories.cdlib.org/are_ucb/1027) accessed 23 March 2009.

WHO (1978). www.who.int/hpr/NPH/docs/declaration_almaata.pdf, accessed 6 February 2008).

World Bank (1993). *World development report.* Washington, DC, World Bank.

4

Current situation in access to drinking-water

Katherine Pond and Stephen Pedley

Any economic appraisal or evaluation is only as good as the physical foundations on which it rests. Describing the physical situation accurately and linking factors causally together convincingly are necessary conditions for putting socioeconomic values on the costs and benefits and arriving at a reliable estimate of the net worth of an intervention. The causal connections in the physical specification are essential in attributing outcomes to the intervention. Attribution is always a challenge in a complex, dynamic situation where many external variables operate. Chapters 5, 6 and 7 are concerned with specifying the physical context for drinking-water interventions. To get an idea of the extent of the problem we are addressing, this chapter provides information on the level of access to safe drinking-water globally and highlights the challenges faced by communities with restricted access to water and those served by small-scale water supply systems.

© 2011 World Health Organization (WHO). *Valuing Water, Valuing Livelihoods*. Edited by John Cameron, Paul Hunter, Paul Jagals and Katherine Pond. Published by IWA Publishing, London, UK.

Part of specifying the physical context of water interventions is to understand the ways in which the target population is behaving and the role that collecting and using water plays in the total scope of activities of people. Many of the technical issues and challenges faced by communities using small-scale systems to supply water are similar, both in the developed and developing country context.

The management and financial aspects associated with small systems often pose specific problems because of their remote and isolated locations. Communication and education measures that are essential in small supply programmes also tend to be hampered for the same reasons.

Improved access to safe water in adequate quantities can enhance health, enrolment in educational programmes, economic productivity and dignity. It thus plays a key role in efforts to reduce poverty – another MDG target. While the majority of MDGs focus on developing countries, developed nations have recognized that small-scale systems to provide community water supplies, particularly for rural, remote and indigenous communities, are the most vulnerable. Such systems are most liable to contamination and failure, and consequently pose a continuous public health risk. Communities relying on small systems to supply drinking-water are often not able to overcome the challenges posed by these systems for a number of reasons:

- Isolation and remoteness lead to increased costs associated with accessing supplies.
- The quality of drinking-water in small systems tends to be poorer, yet the water is sampled less frequently and often not treated.
- The financial resources available for funding capital and operating expenses are limited.
- The per capita costs for water sampling and testing are high.
- Recruiting and training competent or certified operators is a challenge, especially when funding is scarce.
- Little capacity exists to undertake risk assessments or sampling.
- Owners of very small systems and private wells often lack knowledge about or interest in the relationship between poor water quality and ill-health.
- Operators often lack a support network, standard operating procedures and technical support.
- Training for operators and managers of small systems is inadequate and management expertise is lacking.
- The infrastructure of small systems is often characterized by poor construction and inadequate maintenance.
- Communities often lack the skills or financial means to protect water sources or have little influence on factors that may affect water sources.

- The community perception of risk is often inaccurate.
- Risks and risk factors are often hard to quantify and compare in small systems.
- Surveillance of waterborne diseases associated with small systems is especially difficult because of underreporting of waterborne illness and unsystematic collection of data.
- Communities are often faced with a number of other priorities, such as housing, hygiene and socioeconomic problems, which compete for priority with concerns relating to water.
- Communication to the public is deficient, including about the management of water within the home.
- There is insufficient political engagement.
- Regulatory agencies do not have the resources to adequately regulate small systems that provide community water supplies.
- The perception that there is no ownership of a water supply system and no awareness of the true cost of water may result in poor decision-making.
- Poor infrastructure in rural areas in general inhibits delivery of safe water.

To overcome some of the challenges facing small systems that provide community water supplies, the International Network on Small Community Water Supply Management http://www.who.int/water_sanitation_health/dwq/scwsm_network/en/index.html and others have identified the following actions:

- better management of community water supplies;
- management of priorities;
- information generation and dissemination;
- bringing communities together to share experiences;
- development of communication strategies to inform the public and decision-makers about risks;
- development of tools to ensure that decision-makers at community, regional and national levels are aware of their responsibilities;
- advocacy and political will at all levels;
- identification of appropriate regulations for community water supplies;
- commitment and responsibility of governments to investment;
- adequate institutional support to ensure outreach mechanisms;
- capacity-building for water operators and managers, including incentives to stay within the community;
- promotion and strengthening of community-level capacity to manage water supplies, including the establishment of regional networks to facilitate information sharing and mentoring;
- investment by small communities in their own water supply systems.

Cost–benefit or cost–effectiveness analysis can help overcome some of the challenges faced by communities by producing evidence supporting the notion that continued investment in small systems is necessary for community water supplies.

Table 4.1 Annual average commitments to water supply and sanitation by donor

Bilateral or multilateral external support organization	**US$ millions (average)**		
	1999–2000	**2001–2002**	**2003–2004**
Australia	58	21	25
Austria	14	12	19
Belgium	13	37	19
Canada	35	23	80
Denmark	118	31	140
Finland	15	20	9
France	209	176	163
Germany	377	344	366
Greece	1	1	1
Ireland	8	13	18
Italy	45	32	30
Japan	1 159	512	858
Luxembourg	10	12	11
Netherlands	70	155	122
New Zealand	2	2	1
Norway	33	44	18
Portugal	9	1	1
Spain	90	59	63
Sweden	31	51	47
Switzerland	33	29	32
United Kingdom	151	101	52
United States	165	275	521
African Development Fund	37	124	148
Asian Development Bank	50	177	137
European Community	188	193	351
International Development Association	229	675	684
Inter-American Development Bank, Special Operations Fund	54	0	0
UNICEF	34	28	16
Total commitments	3 238	3 147	3 934

Source: www.oecd.org/dac/stats/crs/water

On a global level, among members of the OECD Development Assistance Committee, assistance to water supply and sanitation as a share of total development assistance fell from 8% in 1999–2000 to 6% in 2001–2002 and remained at 6% in 2003–2004. Although there has been a downward trend in the amount of assistance for water and sanitation since the mid-1990s, the trend now appears to be in reverse. But assistance remains concentrated among relatively few donor and recipient countries. Between 2000 and 2004, three quarters of total bilateral support for water supply and sanitation was given by France, Germany, Japan, the Netherlands and the United States. More than half of the allocations were directed to Asia; just 15% went to sub-Saharan Africa (OECD-DAC, 2007). It has been estimated that investment must double annually to achieve the MDGs for the water and sanitation (Global Water Partnership, 2000). Table 4.1 shows donors' commitments to water and sanitation from 1999 to 2004. Most of these resources are used to finance investments in infrastructure (OECD-DAC, 2007). There is evidence that small supplies receive relatively less attention and fewer resources than large supplies.

The new Global Analysis and Assessment of Sanitation and Drinking-water (GLAAS), a UN-Water initiative led by the World Health Organization, will report periodically on policy frameworks, human resources and funding streams for drinking-water and sanitation. Its first report was published in 2010.

As discussed in chapters 7, 9, 10 and 11, the benefits and the costs of improving access to safe water vary considerably depending on the type of intervention selected and the population characteristics. In some cases, for example where vulnerable sub-populations are involved, the benefits of having a drinking-water supply close by may be far greater than for other cases. For policy-makers, health professionals and engineers to make informed choices about the type of intervention to be implemented at a specific locality, it is essential to carry out a sound economic assessment of the various options available in that particular livelihood context. This book therefore does not offer any universal panaceas, but develops a method for economic appraisals and evaluations that can be applied sensibly to a range of local conditions.

GLOBAL LEVELS OF ACCESS TO SAFE DRINKING-WATER

While the specific local contexts differ, there is a growing global challenge in accessing good quality fresh water. Competing interests of agriculture, industry and households, together with growing human populations, continue to place increasing demands on water resources and are having serious consequences for their quality. Globally, approximately one in three people live in areas of moderate to high water stress and it is estimated that two thirds of people will live in a water-stressed area by 2025 (UNEP, 1999). The critical issues in terms of the causes and impacts of lack of water vary by region, but for developing countries the most urgent problem continues to be lack of access to safe drinking-water.

As discussed earlier, problems are not restricted to developing nations. Small communities in all countries face the greatest difficulty in supplying water of adequate quality and quantity because they have small customer bases and therefore often do not have the finances needed to employ experienced managers to maintain and upgrade their water supply facilities. Interruptions in water service because of inadequate management, as well as failures of drinking-water standards, are problems for some of these small systems in both developed and developing nations. Households may respond by storing water at home, but this creates new risks of contamination resulting from inadequate treatment or unsafe storage. In some parts of the world, there is also the risk of increasing the breeding places of mosquito vectors of malaria and dengue. Although the problems of supplying drinking-water through individually operated small systems are well known, the number of small systems has continued to increase in some countries. For example, in the USA approximately 1000 new small systems are constructed each year to provide communities with water (Committee on Small Water Supply Systems, National Research Council, 1997). Currently 9% of Canada's population are on private water systems and 16% on small distribution systems. At least one in ten Europeans (40–50 million people) receive their daily drinking-water from a small or very small supply system, including private wells (Hulsmann, 2005), although the exact figure is not known. In some cases these supplies are not covered by national law and are not monitored unless the owner so requests.

For many people with vulnerable livelihoods, the daily problems associated with access to water seriously deplete energy, health, money and time. Inequalities based on wealth and location, together with flawed policies, mean that poor people pay the most and travel the furthest for water. Achieving even the basic minimum standard of access to water – 20 litres per person per day of safe water from an improved source (WHO, 2003) which can be maintained if the source is within a 30 minute roundtrip from the home – remains a huge challenge (UNDP, 2006).

In 2008, 87% of the global population used improved water sources (WHO/UNICEF, 2010), indicating that if improvements continue the global MDG target will be reached. According to the Joint Monitoring Programme (WHO/UNICEF, 2010) in 2008 some 880 million people were still without access to improved water sources. The global distribution by region is shown in Figure 4.1.

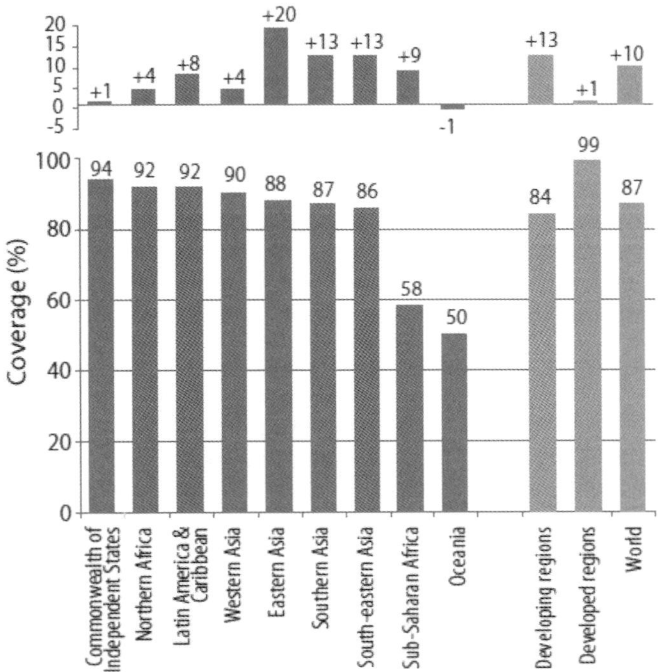

Figure 4.1 Improved drinking-water coverage by region in 2006 and percentage point change 1990–2006. *Source*: WHO/UNICEF (2008)

Each region has shown some improvement since 1990, with eastern Asia showing the greatest improvement in terms of coverage (+20%). The lowest drinking-water coverage is found in sub-Saharan Africa and Oceania. Countries of the CIS, in northern Africa, in Latin America and the Caribbean, and in western Asia have achieved almost 90% coverage or greater (WHO/UNICEF, 2008).

LIVELIHOODS ANALYSIS

In the 1990s, the livelihoods framework was developed to understand how people were coping in challenging situations, notably poor people in rural areas with low agricultural productivity. The framework assumes that every rural society can be seen as having changing patterns of natural, produced, human and social wealth or capital. In summary:

- The physical environment is a reservoir of *natural wealth* important to human well-being in itself and capable of self-development.
- Human activity in the natural environment results in *produced wealth* that has a physical life and productive potential beyond immediate human consumption.
- People can also develop their capabilities into skills whose expression over time as *human wealth* is both a means and an end to long-term development.
- Societies have collective histories of building trust, confidence and mutual security into relationships that constitute *social wealth*.

Conventionally, the basic building block of livelihoods analysis is the household. In the household, significant elements of the wealth portfolio are held in common and used to the joint benefit of all members of the household. In most societies the arrangements for collecting and using drinking-water would be an excellent example of such cooperative activity, though time and energy use in collection may well not be equally shared.

Though the household is the basic building block, livelihoods analysis can be adapted to recognize intra-household inequalities, as in gendered differences in responsibility for collecting drinking-water. It can also be adapted to recognize the importance of numerous forms of migration which give many households a fuzzy boundary in terms of membership. Migration often results in inflows of financial remittances, which can offer possibilities of investing in improving drinking-water access if favourable natural and social conditions exist.

Livelihoods analysis provides a micro-level tool for understanding the opportunities and constraints faced by the target population for an intervention. But it is policy decisions on larger scales, up to and including the global, that are crucial to determining whether resources are made available to radically change access to drinking-water. As discussed in Chapter 1, an increase in the sustainable access to safe water in developing countries is a major international goal that is embodied in the MDGs (United Nations, 2006). The benefit to livelihoods of improved access to safe drinking-water, notably working through improved human wealth (in terms of better health, and increases in time and energy made available for additional productive activities) is a strong argument to support additional resource allocations towards achieving this goal globally.

WATER FOR DRINKING AND LIVELIHOODS IN RURAL AREAS

Sustainable access to sufficient safe drinking-water, sanitary removal of excreta, and personal hygiene are three major factors that contribute to enhancing public health in rural areas. The quality and reliability of a water supply service is important for the improvement of a population's health. Ideally, the whole community should be served efficiently and effectively. However, the water supply service in rural areas often has limited coverage or unreliable continuity. This obliges people to resort to other sources of water or to store water in the household to cover their basic needs. Both of these measures can result in a deterioration of water quality and the consequent exposure of consumers to water-borne pathogens. Unsafe storage can also increase the risk of vector-borne disease transmission.

Data from the WHO/UNICEF Joint Monitoring Programme on Water Supply and Sanitation (WHO/UNICEF, 2010) show that in 2008, 84% of the population without access to an improved source of drinking-water (as defined in Table 4.2) lived in rural areas. Figure 4.2 shows clearly the disparity between urban and rural areas in terms of global populations served by an "improved" water source. This highlights the magnitude of the challenge that faces the international community to improve the living conditions of poor people. According to WHO/UNICEF (2008), in 23 developing countries less than 50% of the rural population have access to drinking-water from improved sources (Figure 4.3), with all resulting health risks.

Table 4.2 Definitions of improved and unimproved sources of drinking-water

Improved water supply	Unimproved water supply
Piped into dwelling, yard or plot	Unprotected dug well
Public tap or standpipe	Unprotected spring
Tube well or borehole	Cart with small tank or drum
Protected dug well	Tanker truck
Protected spring	Surface water (river, dam, lake, pond, stream, canal, irrigation channel)
Rainwater collection	Bottled water

Source: WHO/UNICEF (2010).

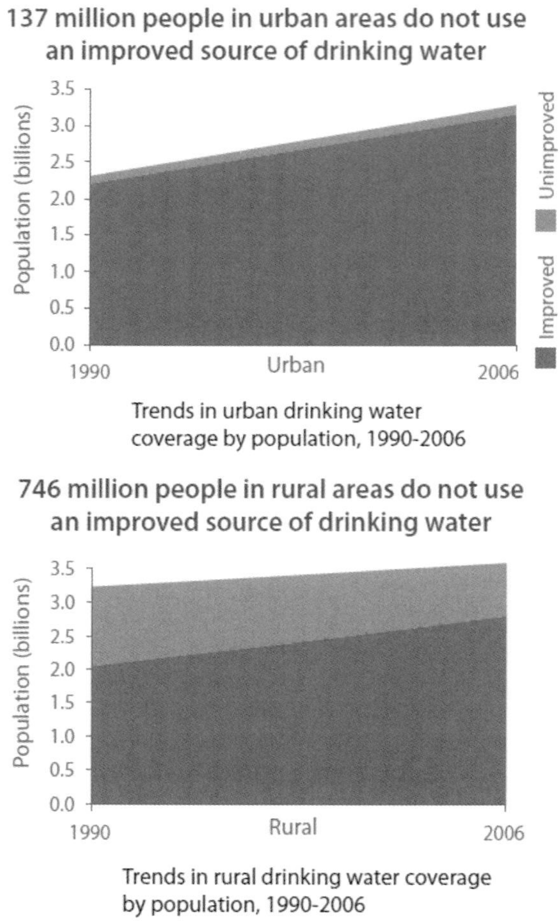

Figure 4.2 Trends in urban and rural drinking-water coverage by population 1990–2006.
Source: WHO/UNICEF, 2008

In its 2006 *Human Development Report*, UNDP (2006) says that the statistics published by the Joint Monitoring Programme may underestimate the numbers of people without access to improved water, because of underreporting by some countries. According to UNDP (2006), three distinctive features of rural water provision explain the low coverage. First, the rural population tend to live in dry areas which are subject to seasonal shortages of rain (a natural wealth constraint in livelihood terms). Second, in most rural areas, people provide, maintain and

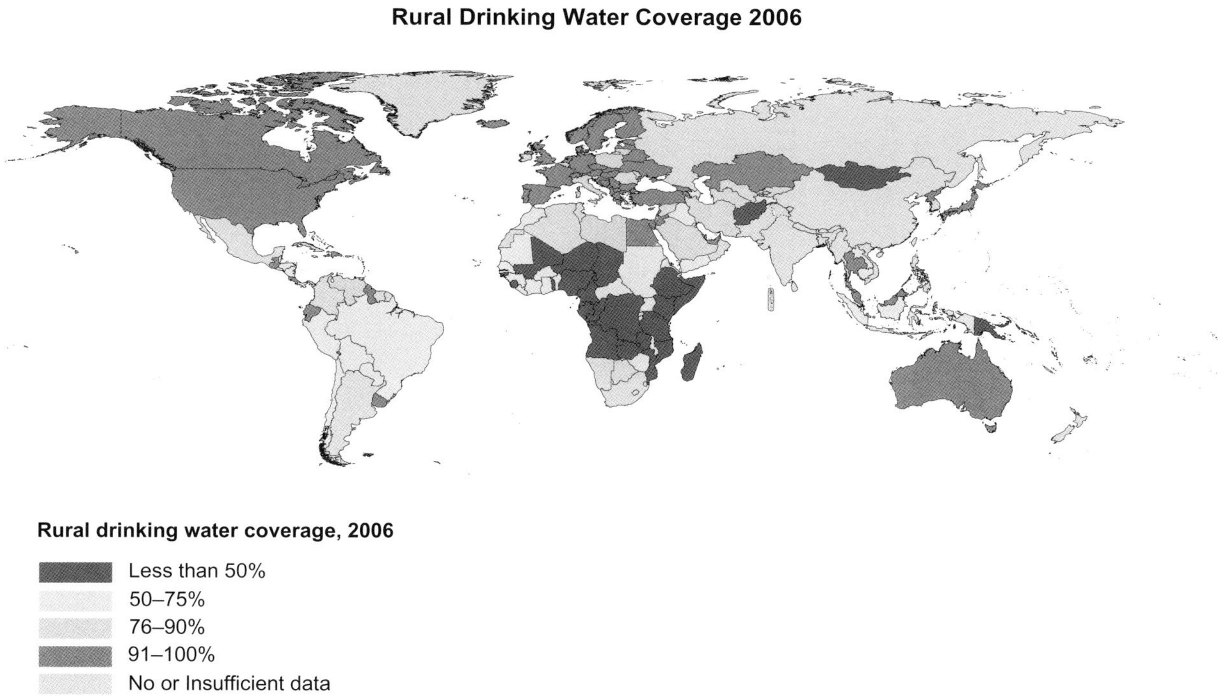

Note: The boundaries and names shown and the designations usedon this map do not imply official endorsement or acceptance by the United Nations.

Figure 4.3 Rural coverage of drinking-water from an improved source in 2006. *Source*: WHO/UNICEF (2008)

expand the water systems themselves, using scarce human and produced wealth resources. This requires high levels of community mobilization (social wealth in livelihood terms) and often the involvement of local government rather than large municipal providers. The accountability of these bodies and the strength of community involvement influences coverage. Third, highly dispersed rural populations, especially in marginal areas, often have very little political influence and therefore little say over resource allocation.

UNDP (2006) also notes that because poor people are particularly vulnerable to the inadequate provision of services, they will benefit disproportionately from improvements to these services. For example, household surveys conducted by UNDP have shown that in Uganda, access to an improved water source reduced the risk of infant mortality by 23%. Similarly, in Egypt, access to a flush toilet reduced the risk of infant death by 57%.

LIVELIHOOD ASPECTS OF ACCESS TO WATER

The health aspects of access to water are well known. Because diarrhoeal diseases are generally of faecal origin, interventions that prevent faecal material entering the environment are likely to be of greatest significance for public health. Interventions with the potential to reduce diarrhoeal disease in rural communities are discussed in Chapter 7. There is less awareness about the livelihood aspects of access to water. Livelihoods comprise the capabilities and assets that people need to make a living and maintain their well-being (UNDP, 2006). In rural areas, water plays a crucial role for obvious reasons. Access to a reliable supply of water allows people to expand their livelihoods, increase productivity and reduce the risks associated with drought.

Most research undertaken in the area of benefits and costs of providing households with sustainable access to safe drinking-water has focused on the relationship between water and disease. Less has been written about the costs to health (other than diarrhoeal diseases) that may affect individuals involved in collecting water. This topic is introduced here and discussed in further detail in Chapter 5.

In broader livelihood terms, women and young girls are particularly at a disadvantage because they sacrifice their time and their education to collect water. Women are also subject to a high degree of physical stress resulting from carrying heavy loads of water. Evidence of such stress and accompanying ill-health is presented using data from a survey conducted in Kibwezi Division of Machakos District, Kenya. Higher-than-average economic dependence on women is shown in the demographic structure. Water collection patterns show that 70% of all trips are made by women over 15 years of age over a median distance of 3.5 km, and that 87% of women collecting water carry loads without any mechanical assistance compared to 42% of men. Using functional criteria, the data suggest a higher prevalence of chronic disablement among women, compared with men, and the disabilities mentioned tend to reflect the hard lifestyle of women in Kibwezi (Ferguson et al., 1986). Previous studies undertaking cost–benefit analysis or cost–effectiveness analysis of improved water availability have tended to ignore health benefits other than reduction in diseases.

A report commissioned by the World Bank concluded that the most obvious benefit is that water is made available closer to rural households (Churchill et al., 1987). This allows time saving, greater convenience, saving of energy and money, and prevention of injury from carrying heavy containers of water.

The time saved by providing access to water close to the home is substantial in some cases. There are still a significant number of areas where water scarcity, at least seasonally, imposes a burden on women in their daily lives. In some parts of Africa, women and children spend eight hours a day collecting water. The proportion of rural women affected by water scarcity is estimated at 55% in Africa, 32% in Asia and 45% in Latin America (United Nations Commission on the Status of Women, 1995). On average, women in developing countries may spend as much as 1.6 hours a day collecting water in the dry season, and 0.6 hours a day in the wet season. Unfortunately, no studies have been done of the relationships between desertification, deforestation and water collection time. A study by Wodon & Blackden (2006) shows how much time women and men spend in collecting water (per trip) in rural areas in selected countries (see Table 4.3).

Table 4.3 Time (in minutes) that women and men spend collecting water (per trip) in rural areas

Benin, 1998		**Ghana, 1998/1999**		**Guinea, 2002/2003**		**Madagascar, 2001**	
Women	**Men**	**Women**	**Men**	**Women**	**Men**	**Women**	**Men**
62	16	44	34	28	6	32	8

Source: Adapted from Wodon & Blackden (2006).

In a study by Roy et al. (2005) in the community of N'atipkong and Ngendui, in Kenya, women reported spending an average of 3.5 hours each day collecting water during the dry season and double that (because hillsides are slippery) in the wet season. They used between 40 litres (elderly women) and 100 litres of water each day. Eight women from this community provided estimates of water quantity collected and time taken (see Table 4.4). The women's answers suggest weekly water collection times of almost 25 hours in the dry season and nearly 50 hours in the wet season. In the community of Kiptagan, where piped water has been introduced, women recalled devoting 13 to 22 hours per week collecting three to four jerrycans of water per day. "Those who are connected to a piped water system," they said "save an average of 15 hours per week. We can now use this time on economic activities."

Table 4.4 Women's domestic water collection times in the dry season in Ngendui, Nyando Basin, Kenya

Woman's identity number	Quantity (litres per day)	Time (hours per load of 20 litres)	Time (hours per day collecting water)
1	80	1	4
2	80	0.5	2
3	60	0.5	1.5
4	100	1	5
5	60	1	3
6	100	1	5
7	40	1	2
8	40	1	2
Average	70	–	3.5

Source: Roy et al. (2005).

Children also collect water, particularly at weekends, but they take longer because they play at the water source, and collect less: 10 litres instead of 20 litres per trip. Nevertheless, the woman or women of the household have less to fetch when the children are involved in collecting water (Swallow et al., 2005).

As a comparison, data from the United Nations Statistics Division (United Nations, 2000) show that water collection times for villages in Kenya average just over 4 hours per day in the dry season, and 2 hours per day in the wet season. The data also indicate times in the range of 4 to 6 hours per day in Botswana, Burkina Faso and Côte d'Ivoire. Water collection times of 17 hours per week are reported for Senegal and 15 hours per week for the dry season in Mozambique. Thus, the water collection times reported for Kiptagan (15 hours per week) and Ngendui (25 to 50 hours per week) are similar to, or higher than, the highest averages reported for Africa.

A study undertaken by Swallow et al. (2005) looked at five villages in the same region of Kenya – the Nyando Basin. Table 4.5 shows that households in these villages spent on average 1.9 hours per day collecting water in the dry season and 44 minutes per day collecting water in the wet season. The average amount of water collected was 100 litres per household per day in the dry season and 25 litres per day in the wet season. It is not clear whether less water is used in the wet season or whether water is obtained from alternative sources.

Table 4.5 Dry and wet season collection of water in five villages in the Nyando Basin, Kenya

	Average	Minimum	Maximum	Mean	Standard deviation
Time spent (hours per day) collecting water during dry season	134	0.02		1.9	1.5
Volume of water (litres) collected in the dry season	139	18	270	100	47
Time spent (hours per day) collecting water during the wet season	93	0	6.0	0.7	0.8
Volume of water (litres) collected per day in the wet season	140	0	160	25	32

Source: Swallow et al. (2005).

A field study undertaken in Kenya, Uganda and the United Republic of Tanzania between 1966 and 1968 (White, Bradley & White, 1972) suggested that the addition of a closer – but still distant – water source would not necessarily increase household water use. White, Bradley & White found that in situations where water must be carried, the quantity brought home varies little for sources between 30 and 1000 metres from the household. The study also showed that the provision of a rural water supply requires a more flexible approach than one that is purely supply driven.

Furthermore, they found that providing a water supply in a rural area promotes greater support for community-based and individual initiatives. In the late 1990s, the study was repeated by Thompson et al. (2001). The study showed that some changes had taken place and discussed the decisions taken by households to deal with the changes. In some places the reliability of the piped water supplies had deteriorated as a result of rising populations and the consequential increasing stress on the system, and because of the lack of capacity and capabilities of government authorities. In such a situation, households may take the decision to store water, as well as look for alternative sources which may be unimproved (and therefore a health risk) or private (and expensive).

Inadequate access to water can restrict a household's choice. In the worst cases, this means a choice between bearing the costs of potential ill-health, using scarce financial resources, or making large expenditures of time and effort (Thompson et al., 2001).

Mertens et al. (1990) found that 10% of women in Sri Lanka had to travel more than one kilometre to their nearest source of water. Feachem et al. (1978) suggest that providing a water source close to the home has very little effect on consumption unless the source is closer than one kilometre (less than 30 minutes roundtrip) from the user's home. However, water consumption doubles or triples when house connections are provided (White, Bradley & White, 1972) and this may significantly improve hygiene practices. It has been suggested that domestic hygiene is the principal determinant of endemic diarrhoeal disease rate (Caincross & Valdmanis, 2006). Therefore, it may be that in some cases the additional cost of a house connection is offset by benefits in time savings. Putting an economic value on the time-saving benefit is discussed in Chapter 11.

In terms of other uses of water, there are also indications of differences reported in the quantity of water used when there is a piped source available compared to when an unpiped source is the only option. Figures 4.4 and 4.5 show the difference in use for amenity (for example, watering gardens) and production (such as consumption by livestock, construction of houses, or irrigation).

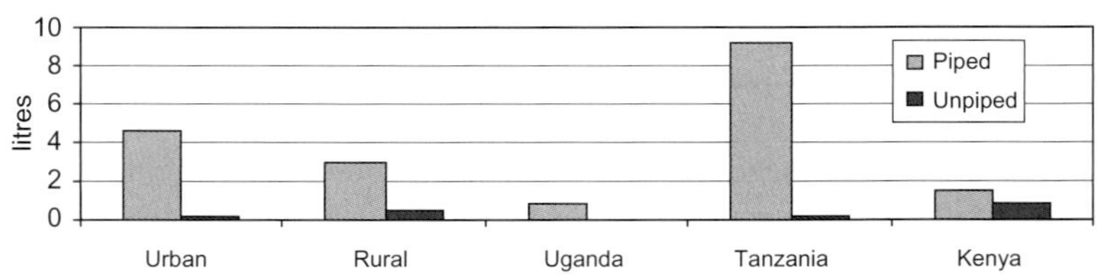

Figure 4.4 Difference in amount of water used (litres) for amenity use when a piped source is available compared to an unpiped source. *Source*: Adapted from Thompson et al. (2001)

As Cairncross & Valdmanis (2006) report, there are several reasons to assign a monetary value to time saving. For example, households often pay others to deliver their water, or pay to collect it from a nearby source rather than collect it free from more distant sources. Thompson et al. (2001) showed that, since the 1970s, the proportion of urban east African households without a piped water supply paying for water had increased from 53% to 80%. Because the poorest households typically spend 90% of their household budget on food, any money that is spent on water will be deducted from the food budget (Cairncross & Kinnear, 1992). Providing cheaper access to water therefore indirectly results in a nutritional benefit to the poorest people.

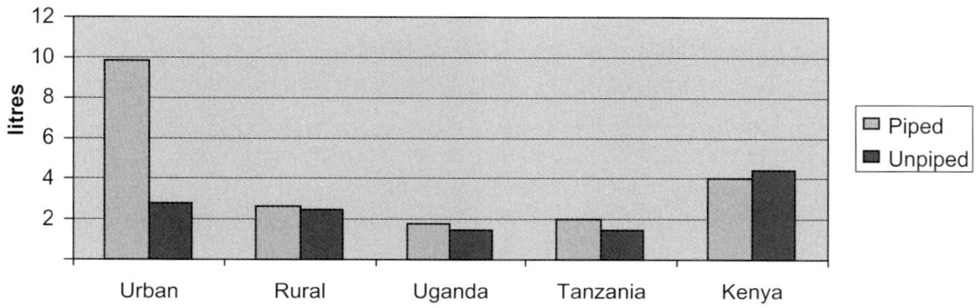

Figure 4.5 Difference in amounts of water used (in litres) for productive uses comparing a piped source with a source that is not piped. *Source*: Adapted from Thompson et al. (2001).

CONCLUSIONS

Small and rural communities are particularly vulnerable to the problems associated with poor access to water and poor quality of small-scale supplies. The availability of a good quality water source close to home has numerous benefits, especially in terms of human wealth, with subsequent linkages to all the other dimensions of livelihoods. Such gains in human wealth have an intrinsic value in terms of quality of life as a developmental end, and as a means for higher economic productivity. In developing countries, women particularly benefit because they are usually the main collectors of water. In developed countries, small water-supply systems are vulnerable to contamination. While generally not facing problems of access to water, communities in developed countries face the challenges related to the quality of the water supply. The benefits associated with access to safe drinking-water provide a strong argument to increase resource allocations to interventions aimed at further improving the current drinking-water situation, as a key entry point for achieving much wider livelihood benefits.

REFERENCES

Cairncross S, Valdmanis V (2006). Water supply, sanitation, and hygiene promotion. In: *Disease control priorities in developing countries*, 2nd ed. New York, Oxford University Press, 0–41.

Cairncross S, Kinnear J (1992). Elasticity of demand for water in Khartoum, Sudan. *Social Science and Medicine*, **34**: 183–189.

Churchill AA et al. (1987). *Rural water supply and sanitation: time for a change.* Washington, DC, World Bank (World Bank Discussion Paper 18).

Committee on Small Water Supply Systems, National Research Council (1997). *Safe water from every tap: improving water service to small communities*. Washington, DC,Water Science and Technology Board, Commission on Geosciences, Environment, and Resources, National Research Council, National Academy Press.

Feachem RG et al. (1978). Sanitation and disease: water, health, and development–an interdisciplinary evaluation. London, Tri-Med Books.

Ferguson A, Praag E van, Absalom EO (1986) *Kibwezi health risk study*. Nairobi, African Medical and Research Foundation.

Global Water Partnership (2000). *Towards water security: a framework for action.* Stockholm and London, Global Water Partnership.

Hulsmann A (2005). Small systems large problems: a European inventory of small water systems and associated problems. WEKNOW, The Netherlands,

Mertens TE et al. (1990). Determinants of water quality, availability, and use in Kurunegala, Sri Lanka. *Tropical Medicine and Parasitology*, 41:89–97.

OECD-DAC (2007). *Has the downward trend in aid for water reversed? Measuring aid for water*. Paris, Organisation for Economic Co-operation and Development (www.oecd.org/dac/stats/crs/water, accessed 31 March 2008).

Roy J et al. (2005). *Getting access to adequate water: community organizing, women and social change in western Kenya.* International Workshop on African Water Laws: Pluralistic Legislative Frameworks for Rural Water Management, 26–28 January 2005, Johannesburg(http://www.nri.org/waterlaw/AWLworkshop/ROY-J.pdf, accessed 22 June 2007).

Swallow B et al. (2005). *Dynamics of poverty, livelihoods and property rights in the lower Nyando basin of Kenya*. Paper presented at the International Workshop on African Water Laws: Pluralistic Legislative Frameworks for Rural Water Management in Africa, 26–28 January 2005, Gautang, South Africa.

Thompson J et al. (2001). Drawers of water II: 30 years of change in domestic water use and environmental health in east Africa. London, International Institute for Environment and Development.

United Nations (2000). *The world's women 2000: trends and statistics*. New York, United Nations Statistics Division, United Nations Publications.

United Nations (2006). *2006 Millennium Development Goals report*. New York, United Nations (http://mdgs.un.org/unsd/mdg/Resources/Static/Products/Progress2006/MDGReport2006.pdf, accessed 28 July, 2008).

United Nations (2007) Earth Summit +5. Special Session of the General Assembly to review and appraise the implementation of Agenda 21, New York, 23–27 June 1997. (http://www.un.org/ecosocdev/geninfo/sustdev/womensus.htm, accessed 7 May 2008).

United Nations Commission on the Status of Women (1995). Preparations for the Fourth World Conference on Women: action for equality, development and peace–review and appraisal of the implementation of the Nairobi forward-looking strategies for the advancement of women. Second review and appraisal of the implementation of the Nairobi Forward-looking Strategies for the Advancement of Women. Report of the Secretary-General. New York, United Nations. http://www.un.org/documents/ecosoc/cn6/1995/ecn61995-3add1.htm, accessed 25 July 2008.

UNDP (2006). Human Development Report: beyond scarcity – power, poverty and the global water crisis. New York, United Nations Development Programme.

UNEP (1999). *Global environment outlook 2000*. Nairobi, Earthscan publications (http://www.unep.org/geo2000/, accessed 25 March 2009).

WHO (2003). *World Health Report.* Geneva, World Health Organization.

WHO/UNICEF (2006). *Meeting the MDG drinking-water and sanitation target: the urban and rural challenge of the decade*. Geneva/New York, World Health Organization and United Nations Children's Fund.

WHO/UNICEF (2008). *Progress on Drinking-water and Sanitation, special focus on sanitation.* New York/Geneva, United Nations Children's Fund and World Health Organization.

WHO/UNICEF (2010). *Progress on Sanitation and Drinking-water, 2010 update*. Report of the WHO/UNICEF Joint Monitoring Programme. Geneva/New York, World Health Organization/United Nations Children's Fund.

White GF, Bradley DJ, White AU (1972). *Drawers of water: domestic water use in East Africa*. Chicago, University of Chicago Press.

Wodon Q, Blackden CM (2006). *Gender, time use, and poverty in sub-Saharan Africa*. Washington, DC, World Bank (World Bank Working Paper 73).

5

Defining the current situation – epidemiology

Paul R Hunter and Helen Risebro

The first step in any economic appraisal or evaluation is to understand the underlying problem being addressed (see Chapter 1). Clearly, such an analysis of drinking-water interventions will have a strong public health element. This chapter discusses the role of epidemiology in identifying the burden of disease[1] in a community that may be attributable to lack of access to safe drinking-water or adequate sanitation.

In order to determine the scale of the problem, there are three questions to be asked:

- What is the burden of disease in the target group?

[1] WHO measures the burden of disease in disability-adjusted life years (DALYs). This time-bound measurement combines years of life lost as a result of premature death and years of healthy life lost because of time lived in a status of less than full health. Mortality and morbidity are linked to other indicators such as financial costs.

© 2011 World Health Organization (WHO). *Valuing Water, Valuing Livelihoods*. Edited by John Cameron, Paul Hunter, Paul Jagals and Katherine Pond. Published by IWA Publishing, London, UK.

- What proportion of the burden of disease is caused by deficiencies in access to drinking-water that are to be remedied by the intervention?
- Are there any spin-off livelihood effects that would result from the outcomes of the intervention?

This chapter focuses on the first two questions. Specific data challenges related to livelihood analysis are raised in Chapter 6.

This chapter aims to assist decision-makers in gathering evidence to enable them to make an informed decision about whether or not there is a public health need for an intervention. A decision about the existence of a public health need is a prerequisite to undertaking a full economic assessment. The chapter outlines some of the methods of epidemiology, as a basis for better understanding epidemiological papers and reports. The chapter then goes on to describe how existing analyses may be used to estimate disease burden.

MEASURES OF DISEASE OCCURRENCE

The two predominant measures of disease occurrence are prevalence and incidence.

Prevalence measures the amount of disease in a population at a given time and can be expressed as a percentage or shown as cases per population:

$$\frac{\text{Number of existing cases in a defined population at a given point in time}}{\text{Number of people in the defined population at the same point in time}}$$

The *point prevalence* is a single assessment at a fixed point in time, whereas the *period prevalence* is the percentage of a population who have the disease at any time within a stated period. Period prevalence is preferred in infectious disease epidemiology because it can be used when there are repeated or continual assessments of the same individuals over a period of time (such as multiple episodes of diarrhoea).

Longitudinal prevalence can be calculated using the following formula (Morris et al., 1996):

$$\frac{\text{Number of days with diarrhoea}}{\text{Number of days under observation}}$$

Incidence measures the number of new cases of disease in a population over a specific time. When the population is constant, the *incidence risk* is measured as:

$$\frac{\text{Number of people who develop disease over a defined period of time}}{\text{Number of disease-free people in that population at the start of the time period}}$$

When the population is not constant, for example, through deaths, migration, births, or through additional participant recruitment, the *incidence rate* should be calculated as:

$$\frac{\text{Number of new events in a defined population over a defined period of time}}{\text{Total person}-\text{time at risk during the defined period of time}}$$

When studying illnesses that last a short time (days or a few weeks), such as acute diarrhoea, then incidence would usually be the most appropriate measure. For more protracted diseases, such as the health effects of arsenic poisoning, prevalence would be the more appropriate measure.

ESTIMATING DISEASE OCCURRENCE

There are different approaches to estimating disease occurrence in a population. The choice of approach will depend on many different factors, such as the amount of resources available and the accuracy of result required. Whatever approach is used, one of the most important starting points is to develop a case definition.

Case definition

The case definition is essential for both the epidemiological studies and any subsequent cost–benefit analyses. The case definition will enable the researcher to know whether or not a particular health event should be included in the analysis and will enable the cost–benefit analyst to determine the cost of the disease outcome. A case definition may be based on symptoms (such as the presence of diarrhoea or clinical features of arsenic poisoning) or the results of laboratory investigations (such as whether or not a stool sample is positive for *Cryptosporidium*). For example, WHO defines diarrhoea as three or more loose or fluid stools (which take the shape of a container) in a 24-hour period (WHO, 1993). Case definitions may also include age ranges, geographical location or dates of onset.

Whatever case definition is used, it should be clear and standardized to minimize disease misclassification bias. Standardizing case definitions is especially important when there is more than one field researcher or interviewer or clinician, or when the study is carried out in more than one community. This is because definitions of diarrhoea can be culture- or person-specific. For example, a study conducted in a rural municipality in Nicaragua in Central America identified a classification encompassing nine different types of diarrhoea (Davey-Smith et al., 1993). The classifications used in Nicaragua were influenced by the place and the person consulted for treatment. The source of any existing data on the use of health care should therefore always be carefully considered.

Primary surveys

Where prior information is not available from local health care facilities or is suspected to be unreliable it may be most appropriate to collect data directly from the population concerned. Such primary data are especially valuable for estimating the burden of disease for illnesses that are unlikely to cause people to visit their local health care provider. In particular, such data collecting is valuable for self-reported diarrhoea. Data collecting can also be especially valuable in poor or remote communities, with limited access to health care. In these circumstances, even people with severe and chronic disease may not come to the attention of the health services. Such population surveys are, however, poor at identifying uncommon illnesses. These surveys usually involve a questionnaire and this may solely be concerned with determining whether the respondent reports various symptoms, to enable a diagnosis on whether or not the symptoms satisfy the case definition. Sometimes a physical examination, or even a laboratory or radiological examination may be included. For example, stool samples may be collected in a study of gastrointestinal disease. Examination of the teeth and radiological examination of the skeleton may be necessary for exposure to fluoride at toxic levels.

There are two forms of population survey: the cross-sectional survey and the cohort study.

Cross-sectional studies are a relatively quick way of getting an estimate of disease incidence or prevalence in a community. Cross-sectional studies look at the disease status of all or a sample of a population at a particular moment in time. In general, each individual would be contacted only once. For diseases with seasonal variation in their incidence, the results of the survey would clearly depend on what time of year the study was carried out. Conducting repeat studies or lengthening the duration of the data collection period may improve the results. Cross-sectional studies can be conducted using various ways of contacting the participants. The choice of approach will depend on resources available, costs and existing communications. Researchers may contact people by visiting them in their homes, or by post or telephone. Response rates are generally poorest for postal surveys (typically around 50%), slightly better for telephone surveys and best for direct visits to the home. Clearly, if a physical examination is part of the survey, face-to-face contact is essential.

A cohort study follows a group of individuals over a period of time. During this period, researchers monitor participants for the appearance of the disease outcome of interest. Usually the initial contact includes an expanded baseline questionnaire. There are several ways of recording the presence or absence of illness. Probably the simplest method is for the researcher to visit or contact the participant at regular intervals. Sometimes people are asked to keep a daily diary of symptoms, which is then collected by the researcher. An example of a pictorial diary used for recording frequency and consistency of stool is shown in Figure 5.1.

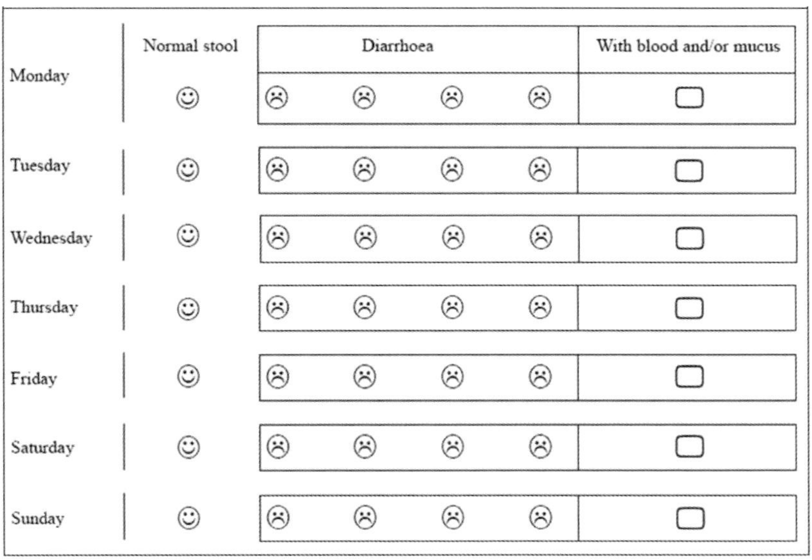

Figure 5.1 Example of a symptom diary. *Source*: Wright et al. (2006).

Some of the advantages and disadvantages of cross-sectional and cohort studies are listed in Table 5.1. In general, cross-sectional surveys are quicker and less costly than cohort studies but they can suffer from recall bias in that people may overreport very recent or current diarrhoea (Boerma et al., 1991). Cohort studies are generally more expensive and take longer, but they are better at detecting the risk factors that predict illness. However, cohort studies seem to suffer from a fall-off in enthusiasm for reporting (respondent fatigue) which could lead to an underestimate of actual disease burden (Strickland et al., 2006; Verbrugge, 1980).

Table 5.1 Advantages and disadvantages of the cross-sectional and cohort design for estimating disease occurrence

Study design	Advantages	Disadvantages
Cross-sectional	• Quick and relatively easy as no follow-up is required • Can be used to investigate many exposures and outcomes • Useful for measuring the true burden of disease in a population	• Problem with direction of causality (reverse causality) • In etiological studies, survival bias a potential problem as the sample is based on prevalent rather than incident cases • Not efficient for rare diseases • Recall bias • Not suitable for diseases of short duration • Prevalence estimates are potentially biased by low response rates • Migration in and out of population influences prevalence
Cohort	• Direct measurement of incidence of disease in exposed and unexposed groups • Time relationships between exposure and disease known • Reduced bias in exposure measurement • Multiple outcomes can be studied • Allows direct calculation of attributable risk • Effects of rare exposures can be evaluated by careful selection of cohorts • Natural history of disease can be evaluated	• Can be expensive and time-consuming • Not useful for rare diseases • Historical cohorts are very dependent on quality of records • Losses of follow-up can bias findings (selection bias) • Outcome assessment can be influenced by knowledge of exposure (information bias)

Source: Adapted from Bowling & Ebrahim (2005).

Sampling

In the studies described so far it is unlikely that the analysts will have sufficient resources to survey the entire population. What they will have to do is sample a proportion of the actual population. Sampling is a method of selecting units (e.g., individuals, households etc) from a defined population (e.g. villages, towns, countries), with the aim of using this subset to make inferences about the population as a whole. One of the problems with sampling is that it can lead to bias. For example, if only people from the wealthier communities are sampled, disease incidence may be underestimated. In order to avoid systematic sampling errors, it is advisable to select a random sample where each sampling unit (individual, household, family etc.) should have an equal probability of selection. As shown in Table 5.2, there are various approaches to random sampling.

The choice of sampling method will depend upon a number of factors, including time, resources and study design. A combination of sampling strategies may be employed. For example, in a study conducted in the Democratic Republic of the Congo, Mock et al. (1993) selected an equal number of villages from four provinces (stratified cluster sampling). Subsequently, a systematic sample of women with children under the age of 30 months was selected from each village. In Zimbabwe, Waterkeyn & Cairncross (2005) took a representative systematic sample of 25 health clubs from each of two districts. Three members of each club (the 10th, 20th and 30th members on the club register) were then selected for inclusion in the study. In addition, 4 neighbours of each of the three chosen club members were included giving a total of 15 participants clustered around each health club (cluster sample).

The sample chosen for analysis should be representative of the population under study and large enough to make statistical inference. Where prevalence of a condition is low, a larger sample is required. If a sample is too small, there will be a lack of precision in results. Specific statistical methods exist for calculating required sample size; an array of textbooks and journal articles cover this in more detail (see Woodward, 2005; Kirkwood & Sterne, 2006). It is advisable to contact a statistician before selecting a sample and designing the survey or investigation.

Using existing local health data

A less costly and usually more rapid way of determining disease occurrence is to use existing data collected by local health services. Ethical principles apply to such data, especially when it is possible to identify which individual relates to each report. The big problem with these datasets is that they are based only on people who are ill enough or have sufficient resources to present themselves to the health services. Surveillance may not even capture all people who present to health care. For example, if surveillance is based on positive results from laboratory tests, then patients who do not have samples taken will not be included in the analysis. The cumulative loss of data on cases (because people do not seek medical care, or because no tests are done even for those who seek care and there are thus no laboratory reports, or because diagnoses are not reported) is known as the reporting pyramid (Wheeler et al., 1999; O'Brien & Halder, 2007).

Table 5.2 Approaches to sample selection

Sampling method	Description
Simple random sampling	Each individual or unit is numbered and a sample of the required size is selected using a random numbers table or a computer-generated list of random numbers.
Systematic random sampling	A system is in place to select a smaller sample from a larger population. Sampling may begin at a random point but all individuals or units thereafter follow in sequence. If the list is ordered in a particular way, this can lead to serious bias. If the list is ordered in a random way, this is called *quasi-random sampling*.
Stratified random sampling	The inventory of people is divided into specific subgroups (strata) of, for example, age and sex categories, and a random sample is drawn from each. This sampling method may be practised when a specific subgroup is over-represented and an even distribution of people in strata is required.
Cluster random sampling	The population is divided into sub-populations (clusters) and a random sample of these clusters is selected (this differs from stratified sampling where all subgroups are selected). For example, to minimize travel costs and time, it may be appropriate to randomly select a number of villages from within a rural area.
Multistage random sampling	A random sample of subgroups is drawn from the target population; further random samples can be drawn from these subgroups.
Sampling for telephone interviews	A list of numbers in the study area is obtained or generated (non-residential numbers should be excluded from the sample) and a random sample is subsequently selected either manually or by means of *random digit dialling*. A less resource intensive method is the use of computer generated lists

Examples of the types of health-care data that may be available include:

- national morbidity and mortality data;
- consultation data:
  - hospital episode and admissions data;
  - emergency department visits (Heffernan et al., 2004);
  - general practice or other health post or clinic consultation data (Boussard et al., 1996);
  - telephone helplines (Rodman, Frost & Jabukowski, 1998; Cooper et al., 2003);
- data from specialist communicable or infectious disease surveillance centres:
  - international (for example, WHO, Enter-net);
  - national (for example, the Centre for Infection, Health Protection Agency in the United Kingdom, Centers for Disease Control and Prevention (CDC) in the United States, the National Public Health Institute (KTL) in Finland, the National Institute for Public Health and the Environment (RIVM) in the Netherlands);
  - regional or local centres;
- proxy indicators of enteric infectious diseases (EID) and EID outbreaks in the community:
  - pharmaceutical sales or prescriptions for anti-diarrhoeals (Sacks et al., 1986; Beaudeau et al., 1999; Edge et al., 2004);
  - school or work absenteeism records.

Extrapolating from previous studies

It may well be the case that researchers have already conducted studies in the community of interest. Others may have conducted studies that have estimated disease occurrence rates in similar communities to the one of interest. If these studies have been published in the peer-reviewed literature they will be listed on one of the on-line searchable databases such as PubMed (available free of charge), MEDLINE and ISI Web of Science.

When using existing literature it is important to critically assess the quality of the study and the extent to which it can be extrapolated to the community of interest. Recommendations on how to assess the quality of such studies are published elsewhere (Blettner, Heuer & Razum, 2001; Khan et al., 2001; Rushton, 2000; Downs & Black, 1998).

A particularly valuable source of information on diarrhoeal disease incidence in developing countries is provided by the Measure DHS (Demographic and

Health Surveys) Project (http://www.measuredhs.com/). This is a USAID-funded project which has been collecting and analysing data from developing countries since 1984. For many countries there may be reports already available that give estimates of diarrhoeal incidence, albeit only in children under 5 years old.

Using existing global estimates

The most common illness linked to poor access to safe drinking-water is diarrhoeal disease. There have been various attempts to estimate the diarrhoeal disease burden globally (Lopez et al., 2001; Bern, 2004; Prüss & Havelaar, 2001). Diarrhoeal disease is one of the most significant contributors to the preventable disease burden. It is estimated that diarrhoeal disease is the fourth most common cause of death and the second most important contributor to the disease burden globally (Prüss & Havelaar, 2001). There are an estimated 2.5 million deaths and almost 100 million DALYs lost from diarrhoeal disease per year (DALYs are explained in Chapter 9). The burden of disease resulting from diarrhoeal illness falls most heavily on the youngest, especially children under 5 years old. Diarrhoeal disease also predominantly affects the poorest countries, and the poorest communities within those countries. In contrast, in developed countries, diarrhoea is not even in the top 10 causes of disease burden (Prüss & Havelaar, 2001). Estimated mortality and morbidity from diarrhoeal disease, and rural population with unimproved water supply are shown, by country, see Annex.

ESTIMATING DISEASE ATTRIBUTABLE TO A SPECIFIC ENVIRONMENTAL RISK

Once studies, as described above, have shown that a particular disease is an important public health problem in the community, the next step is to identify the important environmental risk factors that are driving the occurrence of that disease. There are three approaches to estimating the contribution of a risk factor: some form of quantitative risk assessment; one or more epidemiological investigations; or the use of pre-existing global estimates. Furthermore, the DHS data sources referred to above can be used to estimate the disease burden attributable to poor water and sanitation.

Risk assessment

Quantitative microbial risk assessment has become a popular tool in recent years, especially in North America. Quantitative microbial risk assessment uses existing data about the infectivity and distribution of pathogenic microorganisms or indicator bacteria to estimate risk to human health. The four stages of a quantitative microbial risk assessment are:

- hazard identification;
- exposure assessment;
- dose–response analysis;
- risk characterization.

An accessible review of quantitative microbial risk assessment methods is given elsewhere (Hunter et al., 2003). Howard, Pedley & Tibatemwa (2006) give a good example of its use in small rural settings in developing countries.

Quantitative microbial risk assessment has advantages over epidemiological studies where the disease under investigation is uncommon, where the costs of an epidemiological study would be too great or where serious time constraints apply. A disadvantage is that the input variables and especially the exposure values may not be known with any degree of accuracy and so there may be large uncertainty around the results.

It is highly unlikely that the analyst will need to generate new dose–response data, as there will be plenty of studies already reported that have this information. For some chemical agents, such as arsenic, there are reasonably well-defined relationships between concentration of arsenic in drinking water and risk of disease (WHO, 2001). The principal problem in studying arsenic is that there are different disease outcomes, such as skin lesions and cardiovascular disease, each of which have different dose–response curves. A general problem with microbiological dose–response curves is that these curves have had to be extrapolated, often from minimal data, and so carry with them significant uncertainty. Another problem with microbiological dose–response curves is that they are usually pathogen-specific and so require data on or estimations of the concentrations of multiple pathogens likely to be present in the drinking-water options under investigation.

Unlike dose–response data, exposure data will often be specific to the community under investigation. However, national or local authorities such as health agencies, administrative bodies or water and sewerage utilities often undertake routine analyses of water quality in their jurisdictions. Even if routine datasets are not available there may well be previous studies that have collected water samples for analysis in the community of interest or in other similar communities.

Primary data collection would include collection of data on basic microbial or chemical analyses, such as *E. coli* counts or arsenic concentrations. Primary data collection for risk assessment would also include basic sanitary surveys.

It is unlikely that sampling for specific pathogens such as *Cryptosporidium* or noroviruses would be available. Hunter et al. (2003) and Howard, Pedley & Tibatemwa (2006) provide further information that will be of help to analysts, particularly on how to conduct the risk characterization.

Epidemiological approaches

Several epidemiological methods are available to estimate the contributions of specific risk factors or transmission pathways to disease burden. These include ecological studies, case-control studies, cohort studies and prospective studies. In regard to waterborne disease, the advantages and disadvantages of these different methods have been discussed in detail elsewhere (Hunter, Waite & Ronchi, 2002). There are also many introductory textbooks on epidemiology, and consequently this discussion will not describe epidemiological methods in detail.

One of the easiest epidemiological approaches is the ecological study. In ecological studies, the unit of observation and analysis is at the group, population or community level, rather than at the level of the individual. Frequently, the data used for observation and analysis are derived from existing data sources (secondary data). The Geographical Information System (GIS) described in Box 5.1 can be very useful in compiling and analysing this type of data. One of the simplest ecological approaches is to estimate disease incidence in areas with high and low exposure to a potential risk factor such as poor quality water. However, such studies may be susceptible to significant confounding. For example, in a study comparing a poor community with no sanitation to a wealthy community with well-managed sanitation, it will be difficult to answer the question of how much of the difference in illness is related to sanitation, because of the many confounding factors linked to wealth.

Cross-sectional studies, as discussed above for the collection of disease incidence data, can also be useful for collecting data on potential risk factors.

In case-control studies, people with an illness are interviewed about past exposure to possible risk factors. The same questions are also asked of controls (people without the disease). Assumptions about the importance of particular risk factors are then made, based on statistical analyses of the proportion of

Box 5.1 Uses of the Geographic Information System

The Geographic Information System (GIS) is a computer based graphics software program which enables the user to capture, store, manipulate, analyse and display spatially referenced data from a number of sources. Separate data sources based on geographically referenced information can be connected via relational databases and used to display any number of data attributes in the form of maps. The integrated information can be used in local and regional resource and environmental planning as well as spatial studies of infectious disease. For example, in Germany GIS has been applied to spatial patterns of diarrhoeal illness in relation to groundwater and surface water supplies (Dangendorf et al., 2003). In Nigeria, GIS was used to evaluate the health impact (diarrhoeal illnesses) of 39 separate water sources displayed in terms of layers, such as hydrology, geology and environmental pollution (Njemanze et al., 1999).

Examples of GIS software include EDINA Digimap (for example, through which the United Kingdom Ordnance Survey Data Collection can be accessed) and ESRI ArcGIS 9.1 (http://www.esri.com). The University of Edinburgh hosts a site with a comprehensive index of GIS resources: http://www.geo.ed.ac.uk/home/giswww.html.

It should be noted that when data are gathered from diverse geographical areas, it may be difficult to compile the data for a specific small area without complex analysis. This can lead to difficulties of interpretation.

cases and controls reporting exposure. Case-control studies are valuable for investigating multiple risk factors. However, they usually focus on a single disease, and it can be difficult to use the output from these studies to estimate disease burden attributable to a single transmission pathway.

In contrast, cohort studies follow people with different levels of exposure (say those with and without access to improved sanitation) and observe how much disease develops in each group. The big advantage is that many different disease outcomes can be observed. However, cohort studies are more costly than case-control studies, because people are followed over time.

An intervention may comprise new sanitation facilities, improved access to water, a change in the water treatment process, or the introduction of educational or behavioural programmes. Intervention studies can be carried out using a variety of study designs including cross-sectional, cohort and randomized controlled trial. Intervention studies can be conducted under natural conditions (accidental trials, such as outbreaks), under uncontrolled conditions (public measures, such as the introduction of a new water treatment plant), or under controlled conditions (clinical trials or field studies) (Payment & Hunter, 2003).

The randomized controlled trial is one of the most robust epidemiological study designs. It permits simultaneous comparison of outcomes in a group of individuals. Study participants are randomly assigned to one or more intervention groups (where the intervention is expected to influence disease status), or to the control group (which receives either the status quo or a placebo – sham – intervention). A comprehensive review of randomized and quasi-randomized controlled trials assessing diarrhoeal disease outcomes according to types of water quality intervention can be found in Clasen et al. (2006).

Existing global and regional estimates

In the absence of local data, or data from similar situations, use can be made of global estimates by determining what water and sanitation scenarios communities fit into. This will then give rough estimates of the disease risk attributable to inadequate water, sanitation and hygiene. There have been a number of attempts to estimate the global burden of disease that may be attributable to lack of access to safe drinking-water and to adequate sanitation, and to poor hygiene (Prüss et al., 2002). Prüss et al. estimated that globally these factors are responsible for 4.0% of all deaths and 5.7% of disease burden. They then went on to define different scenarios of water and sanitation provision, and estimated deaths and attributable disease burden for each (these scenarios are presented in Table 5.3).

Scenario I represents the minimum theoretical risk, namely no disease transmission through unsafe water, sanitation and hygiene; scenario II is the situation typically encountered in developed countries. These scenarios have very low to medium loads of faecal-oral pathogens in the environment, characterized by more than 98% coverage by improved water supply and sanitation. Scenarios III–VI are based on a high faecal-oral pathogen environment, typical for developing countries with less advanced water and sanitation provision. Scenario III represents piped water in-house and improved sanitation, but this scenario does not occur widely.

NONCOMMUNICABLE DISEASE

In contrast to estimates for diseases of microbiological origin, burden of disease estimates attributable to chemical contamination are less well developed at a global level. A wide range of chemical and radiological contaminants of drinking-water have been implicated in human disease (Hunter, 1997). Although toxicological assessments do not have to deal with the problems of prior immunity, there are considerable difficulties in assessing disease burden. There

Table 5.3 Exposure scenarios

Scenario	Description	Environmental faecal-oral pathogen load	Minimum and realistic risk of diarrhoeal disease relative to scenario I
VI	Population not served with improved water supply and no improved sanitation in countries that are not extensively covered by those services (defined as less than 98% coverage), and where water supply is not likely to be routinely controlled	Very high	6.1/11.0
V(ii)[a]	Population having access to improved water supply but not served with improved sanitation in countries that are not extensively covered by those services (defined as less than 98% coverage), and where water supply is not likely to be routinely controlled	Very high	4.9/8.7
V(i)[a]	Population having access to improved sanitation but no improved water supply in countries where less than 98% of the population is served by water supply and sanitation services, and where water supply is likely not to be routinely controlled.	High	3.8/6.9
IV	Population having access to improved water supply and improved sanitation in countries where less than 98% of the population is served by water supply and sanitation services and where water supply is likely not to be routinely controlled.	High	3.8/6.9
III	Piped water in-house and improved sanitation services in countries where less than 98% of the population is served by water supply and sanitation services, and where water supply is likely not to be routinely controlled.	High	2.5/4.5
II	Population having access to piped water in-house where more than 98% of the population is served by those services; generally corresponds to regulated water supply and full sanitation coverage, with partial treatment of sewage, and is typical in developed countries.	Medium to low	2.5/2.5
I	Ideal situation, corresponding to the absence of transmission of diarrhoeal disease through water, sanitation or hygiene	Very low	1/1

[a]Transitions between exposure levels V(i) and V(ii) do not generally occur.
Source: Adapted from Prüss et al. (2002) and Prüss-Üstün et al. (2004).

is still considerable debate over whether or not many chemicals in water actually cause disease. For example, the question of whether aluminium and dementia are linked remains open (Hunter, 1997; Flaten, 2001). Even where the association is accepted, the balance of acute disease and latent or chronic disease may be far from clear. Recent considerations of burden of disease attributable to chemical contaminants of water have included nitrate (Fewtrell, 2004), arsenic (Fewtrell, Fuge & Kay, 2005) and fluoride (Fewtrell et al., 2006).

Musculoskeletal disease makes a significant contribution to global disease burden, and causes substantial disability in both developed and developing countries (Brooks, 2006). How much of that burden of disease is attributable to carrying water is unclear. There is uncertainty both over the total disease burden and the proportion attributable to carrying water. It is likely that the major health effect of carrying water is low back pain, but global estimates of disease burden attributable to back pain are lacking (Brooks, 2006). We are not aware of any good studies directly linking the carrying of water and back pain. When such information is lacking then the assessor may have to extrapolate from studies of similar exposures. For example, Moore, White & Moore (2007) found that children who carry heavy backpacks to school may suffer from increased musculoskeletal symptoms. In the absence of any usable study, the assessor may have to fall back on a process of soliciting expert opinion.

Noncommunicable diseases pose a number of challenges for conducting an economic analysis. As already discussed, existing epidemiological evidence is often relatively poor compared to microbiological data. Also, more so than for infectious diseases, the actual disease burden can vary substantially, even between very similar communities situated relatively close to each other, because of marked differences in contaminant concentrations (Rahman et al., 2005). For some contaminants, drinking-water (either through consumption or through cooking) is the only or predominant source of exposure, for example arsenic (Fewtrell, Fuge & Kay, 2005). For others, such as lead, water is one of the several possible routes of exposure (Romieu et al., 1994).

In these cases it may be difficult to determine the disease burden attributable to water exposure. Reliance on national or regional estimates is also likely to be problematic. These problems with assessing the noncommunicable disease burden are probably responsible for the general exclusion of noncommunicable diseases from earlier economic analyses. However, given that many of these illnesses would be chronic and may be common in affected communities, their contribution to disease burden is likely to be substantial. It is probable that the exclusion of noncommunicable diseases from analyses would, in certain communities, heavily understate the benefits of improving the water supply.

CONCLUSION

This chapter has considered different approaches to estimating disease burden within a community and then estimating what proportion of that disease may be attributed to a specific environmental risk factor. Some of the approaches would take considerable resources to implement. Usually, the cost–benefit analyst will be able to find previous studies that would suffice for the purpose. Global and regional estimates, though probably lacking precision for any single country, may be enough for most purposes, unless there is evidence of chronic noncommunicable disease.

REFERENCES

Beaudeau P. et al. (1999). A time series study of anti-diarrheal drug sales and tap-water quality. *International Journal of Environmental Health Research.* 9: 293 –312.

Bern C. (2004). Diarrhoeal diseases. In: Murray CJL, Lopez AD, Mathers CD, eds. *Global burden of disease and injury series Volume IV: the global epidemiology of infectious diseases.* Geneva. World Health Organization.

Blettner M., Heuer C. and Razum O (2001). Critical reading of epidemiological papers: a guide. *European Journal of Public Health*, 11: 97–101.

Boerma J.T. et al. (1991). Accuracy and completeness of mothers' recall of diarrhoea occurrence in pre-school children in demographic and health surveys. *International Journal of Epidemiology*, 20: 1073–1080.

Boussard E. et al. (1996). Sentiweb: French communicable disease surveillance on the world wide web. *British Medical Journal*, 313: 1381–1382.

Bowling A, Ebrahim S (2005). Handbook of health research methods: investigation, measurement and analysis. Maidenhead, Open University Press.

Brooks PM (2006). The burden of musculoskeletal disease: a global perspective. *Clinical Rheumatology*, 25:778–781.

Clasen T et al. (2006). Interventions to improve water quality for preventing diarrhoea. *Cochrane Database of Systematic Reviews*, Issue 3, Art. No. CD004794. DOI: 10.1002/14651858.CD004794.pub2.

Cooper DL et al. (2003). What can analysis of calls to NHS Direct tell us about the epidemiology of gastrointestinal infections in the community? *Journal of Infection*, 46: 101–105.

Dangendorf F et al. (2003). *The occurrence of pathogens in surface water.* Bonn, University of Bonn, World Health Organization Collaborating Centre (draft report).

Davey Smith G et al. (1993). The cultural construction of childhood diarrhoea in rural Nicaragua: relevance for epidemiology and health promotion. *Social Science and Medicine*, 36: 1613–1624.

Downs SH, Black N (1998). The feasibility of creating a checklist for the assessment of the methodological quality both of randomized and non-randomised studies of health care interventions. *Journal of Epidemiology and Community Health*, 52: 377–384.

Edge VL et al. (2004). Syndromic surveillance of gastrointestinal illness using pharmacy over-the-counter sales: a retrospective study of waterborne outbreaks in Saskatchewan and Ontario. *Canadian Journal of Public Health*, 95: 446–450.

Fewtrell L (2004). Drinking-water nitrate, methemoglobinemia, and global burden of disease: a discussion. *Environmental Health Perspectives*, 112: 1371–1374.

Fewtrell L, Fuge R, Kay D (2005). An estimation of the global burden of disease due to skin lesions caused by arsenic in drinking water. *Journal of Water and Health*, 3: 101–107.

Fewtrell L et al. (2006). An attempt to estimate the global burden of disease due to fluoride in drinking water. *Journal of Water and Health*, 4: 533–542.

Flaten TP (2001). Aluminium as a risk factor in Alzheimer's disease, with emphasis on drinking water. *Brain Research Bulletin*, 55: 187–196.

Heffernan R et al. (2004). Syndromic surveillance in public health practice, New York City. *Emerging Infectious Diseases*, 10: 858–864.

Howard G, Pedley S, Tibatemwa S (2006). Quantitative microbial risk assessment to estimate health risks attributable to water supply: can the technique be applied in developing countries with limited data? *Journal of Water and Health*, 4: 49–65.

Hunter PR (1997). *Waterborne disease: epidemiology and ecology*. Chichester, Wiley.

Hunter PR et al. (2003). Assessment of risk. In: Ronchi E, Bartram J, eds. *Assessing microbial safety of drinking water: improving approaches and methods.* Paris, OECD/WHO, pages 79–109.

Hunter PR, Waite M, Ronchi E, eds. (2002). *Drinking water and infectious disease: establishing the link.* Boca Raton, CRC Press.

Khan KS et al. (2001). Stage II, phase 5: study quality assessment. In: *Undertaking systematic reviews of research on effectiveness. CRD Report 4* (http://www.york.ac.uk/inst/crd/pdf/crd4_ph5.pdf, accessed 15 March 2007).

Kirkwood BR, Sterne JA (2006). *Essential medical statistics*, 2nd ed. Oxford, Blackwell Science.

Lopez AD et al. (2001). Global and regional burden of disease and risk factors, 2001: systematic analysis of population health data. *Lancet*, 367: 1747–1757.

Mock NB et al. (1993). Socioeconomic, environmental, demographic and behavioral factors associated with occurrence of diarrhea in young children in the Republic of Congo. *Social Science and Medicine*, 36: 807–816.

Moore MJ, White GL, Moore DL (2007). Association of relative backpack weight with reported pain, pain sites, medical utilization, and lost school time in children and adolescents. *Journal of School Health*, 77: 232–239.

Morris SS et al. (1996). Is prevalence of diarrhoea a better predictor of subsequent mortality and weight gain than diarrhoea incidence? *American Journal of Epidemiology*, 144: 582–588.

Njemanze PC et al. (1999). Application of risk analysis and geographic information system technologies to the prevention of diarrheal diseases in Nigeria. *American Journal of Tropical Medicine and Hygiene*, 61: 356–60.

O'Brien SJ, Halder SLS (2007). GI Epidemiology: infection epidemiology and acute gastrointestinal infections. *Alimentary Pharmacology and Therapeutics,* 25: 669–674.

Payment P, Hunter PR (2003). Intervention studies. In: Hunter PR, Waite M, Ronchi E, eds. *Drinking water and infectious disease: establishing the links*. Boca Raton, CRC Press.

Prüss A, Havelaar A (2001). The global burden of disease study and applications in water, sanitation and hygiene. In: Fewtrell L, Bartram J, eds. *Water quality: guidelines, standards and health – risk assessment and management for water-related infectious disease.* London, International Water Association Publishing, 43–59.

Prüss A et al., (2002). Estimating the burden of disease from water, sanitation and hygiene at a global level. *Environmental Health Perspectives*, 110: 538–542.

Rahman MM et al. (2005). Status of groundwater arsenic contamination and human suffering in a Gram Panchayet (cluster of villages) in Murshidabad, one of the nine arsenic affected districts in West Bengal, India. *Journal of Water and Health*, 3: 283–296.

Rodman J, Frost F, Jabukowski W (1998). Using nurse hotline calls for disease surveillance. *Emerging Infectious Diseases*, 4: 329–332.

Romieu I et al. (1994). Sources of lead exposure in Mexico City. *Environmental Health Perspectives*, 102: 384–389.

Rushton L (2000). Reporting of occupational and environmental research: use and misuse of statistical and epidemiological methods. *Occupational and Environmental Medicine*, 57: 1–9.

Sacks JJ et al. (1986). Epidemic campylobacteriosis associated with a community water supply. *American Journal of Public Health*, 76: 424–428.

Strickland MJ et al. (2006). Time-dependent record keeping fatigue among youth completing health diaries of unintentional injuries. *Journal of Safety Research*, 37: 487–492.

Verbrugge LM (1980). Health diaries. *Medical Care*, 18: 73–95.

Waterkeyn J, Cairncross S (2005). Creating demand for sanitation and hygiene through community health clubs: a cost-effective intervention in two districts in Zimbabwe. *Social Science and Medicine*, 61: 1958–1970.

Wheeler JG et al. (1999). Study of infectious intestinal disease in England: rates in the community, presenting to general practice, and reported to national surveillance. The Infectious Intestinal Disease Study Executive. *British Medical Journal*, 318: 1046–1050.

Woodward M (2005). *Epidemiology: study design and data analysis*. 2nd ed. London, Chapman & Hall and CRC Press.

WHO (1993). The management and prevention of diarrhoea: practical guidance, 3rd ed. Geneva, World Health Organization.

WHO (2001). *Arsenic and arsenic compounds*, 2nd ed. Geneva, World Health Organization (Environmental Health Criteria No. 224). (http://whqlibdoc.who.int/ehc/WHO_EHC_224.pdf, accessed 15 March 2007).

Wright JA et al. (2006). Defining episodes of diarrhoea: results from a three-country study in sub-Saharan Africa. *Journal of Health Population and Nutrition*, 24: 8–16.

Annex Estimated mortality and morbidity from diarrhoeal disease, and rural population with unimproved water supply, by country. Population and water supply data taken from the WHO/UNICEF Joint Monitoring Programme (JMP) for Water Supply and Sanitation (estimates for year 2008) (http://www.wssinfo.org/ accessed 28 march 2011). Diarrhoeal disease data from World Health Organization Department of Measurement and Health Information (estimates for year 2004) (http://www.who.int/healthinfo/global_burden_disease/estimates_country/en/index.html) (Accessed 8 June 2011). DALY and associated death estimates expected to be updated 2012. Death estimates for 2008 available on the above web site.

	Estimated total deaths from diarrhoea, (thousand)	Estimated deaths from diarrhoea per 100 000 population	Age-standardized death rates from diarrhoea per 100 000	Estimated total DALYs attributable to diarrhoea (thousand)	Estimated DALYs attributable to diarrhoea per 100 000 population	Age-standardized DALYs attributable to diarrhoea per 100 000	Population total (thousand)	Total rural population (thousand)	Number rural with unimproved water supply (thousand)	Number rural without piped supply to home (thousand)
Afghanistan	74.58	309.77	202.86	2478	10 291	5289	27 208	20 680	12 615	20 680
Albania	0.16	4.96	5.28	6	182	194	3143	1675	34	301
Algeria	9.24	28.55	41.09	217	670	691	34 373	11 938	2508	5252
Andorra	0.00	0.96	0.40	0	27	31	84	8	0	8
Angola	53.97	345.19	239.57	1777	11 364	6658	18 020	7794	4833	7716
Antigua and Barbuda	0.00	0.90	1.04	0	119	114	86	49	0	49
Argentina	0.38	0.98	0.85	46	121	121	39 882	3182	636	1750
Armenia	0.16	5.34	8.40	7	232	345	3077	1114	78	334
Australia	0.04	0.20	0.11	5	26	30	21 074	2369	0	2369
Austria	0.02	0.26	0.14	2	25	31	8337	2742	0	0
Azerbaijan	2.08	25.01	32.87	75	899	1166	8730	4202	1218	3362
Bahamas	0.00	1.24	1.54	0	124	121	337	55	0	55
Bahrain	0.01	1.58	5.18	1	80	91	775	89	0	89
Bangladesh	73.52	48.84	42.91	2486	1651	1217	160 000	116 688	25 671	116 688
Barbados	0.00	0.83	0.57	0	92	106	255	153	0	153
Belarus	0.02	0.19	0.20	3	29	38	9678	2571	26	720
Belgium	0.09	0.85	0.35	3	27	31	10 590	278	0	0
Belize	0.02	6.31	6.94	1	286	223	300	145	0	56
Benin	10.55	128.33	110.85	336	4084	2660	8662	5094	1579	4992
Bhutan	0.30	48.85	47.74	10	1605	1399	686	449	54	247
Bolivia	3.86	42.86	33.45	133	1474	1007	9694	3332	1100	1766
Bosnia and Herzegovina	0.01	0.13	0.19	1	30	39	3773	1983	40	575
Botswana	1.03	56.62	60.93	33	1799	1436	1921	774	78	503
Brazil	28.92	15.69	14.98	1065	578	532	191 971	27 474	4396	10 440

Brunei Darussalam	0.00	0.29	0.24	0	85	75	392	98	0	98
Bulgaria	0.02	0.23	0.38	2	31	45	7592	2194	0	2194
Burkina Faso	26.75	198.07	141.60	876	6483	3725	15 233	12 257	3432	12 257
Burundi	16.65	220.06	161.35	545	7201	4343	8074	7236	2098	7164
Cambodia	12.12	88.35	83.47	390	2843	2170	14 562	11 425	5027	10 854
Cameroon	20.69	118.82	102.92	643	3691	2436	19 088	8247	4042	8000
Canada	0.58	1.82	0.98	10	31	34	33 259	6526	65	6526
Cape Verde	0.14	27.38	37.20	4	710	588	498	201	36	147
Central African Republic	4.89	118.57	98.26	154	3732	2400	4339	2667	1308	2667
Chad	15.91	162.16	121.30	514	5241	3017	10 913	8008	4485	7928
Chile	0.24	1.46	1.40	18	112	114	16 803	1933	484	1024
China	62.31	4.75	6.17	3642	277	324	1 337 411	760 371	136 867	205 300
Colombia	2.43	5.49	5.63	117	264	239	45 012	11 489	3102	5055
Comoros	0.34	43.66	56.66	10	1276	1045	660	475	14	375
Congo	1.71	48.42	50.11	52	1479	992	3615	1398	923	1356
Cook Islands	0.00	5.23	4.45	0	239	192	19	5	0	5
Costa Rica	0.12	2.86	3.44	6	146	141	4519	1659	149	182
Côte d'Ivoire	25.14	137.54	116.38	814	4455	3018	20 591	10 537	3372	9062
Croatia	0.01	0.33	0.21	1	26	32	4422	1893	56	435
Cuba	0.24	2.11	1.57	12	108	121	11 204	2729	300	1255
Cyprus	0.01	1.36	1.99	1	104	136	862	259	0	0
Czech Republic	0.01	0.09	0.11	2	25	32	10 319	2737	0	246
Democratic People's Republic of Korea	5.19	22.09	26.46	197	837	976	23 818	8903	0	8903
Democratic Republic of the Congo	120.10	211.00	146.37	3940	6922	3860	64 256	42 464	30 574	41 615
Denmark	0.07	1.22	0.61	2	31	35	5458	726	0	0
Djibouti	0.62	78.36	76.30	20	2562	1772	849	107	51	104
Dominica	0.00	0.00	0.00	0	111	105	66	17	0	17
Dominican Republic	1.46	15.63	14.94	50	538	455	9952	3079	492	1416
Ecuador	1.26	9.77	8.40	54	415	343	13 481	4629	556	1203
Egypt	11.57	16.17	15.31	404	565	454	81 527	46 741	935	6076
El Salvador	0.80	12.18	10.61	32	480	385	6133	2413	579	1399
Equatorial Guinea	0.76	159.86	130.02	24	4978	3209	659	400	0	400
Eritrea	3.61	82.87	64.99	122	2797	1657	4926	3909	1681	3909
Estonia	0.00	0.00	0.00	0	25	33	1341	410	12	102
Ethiopia	124.16	161.25	124.82	4003	5199	3219	80 713	67 056	49 622	67 056
Fiji	0.05	5.90	9.45	1	170	169	844	402	0	402
Finland	0.02	0.36	0.15	1	25	30	5304	1947	0	1947

(Continued)

Annex (*Continued*)

	Estimated total deaths from diarrhoea, (thousand)	Estimated deaths from diarrhoea per 100 000 population	Age-standardized death rates from diarrhoea per 100 000	Estimated total DALYs attributable to diarrhoea (thousand)	Estimated DALYs attributable to diarrhoea per 100 000 population	Age-standardized DALYs attributable to diarrhoea per 100 000	Population total (thousand)	Total rural population (thousand)	Number rural with unimproved water supply (thousand)	Number rural without piped supply to home (thousand)
France	0.97	1.59	0.72	20	32	35	62 036	14 061	0	0
Gabon	0.47	37.04	46.19	11	834	723	1448	214	127	193
Gambia	1.21	77.11	74.05	36	2286	1571	1660	721	101	685
Georgia	0.48	10.67	17.06	17	375	597	4307	2037	82	998
Germany	0.92	1.11	0.49	22	27	33	82 264	21 702	0	650
Ghana	22.86	103.66	171.36	455	2063	2031	23 350	11 674	3036	11 324
Greece	0.00	0.00	0.00	3	23	29	11 137	4350	43	43
Grenada	0.01	4.90	5.03	0	196	173	103	71	0	71
Guatemala	4.55	36.70	34.99	140	1129	764	13 686	7045	704	2254
Guinea	10.73	121.53	100.59	339	3841	2400	9833	6454	2517	6389
Guinea-Bissau	3.11	200.91	145.07	101	6497	3689	1575	1106	543	1095
Guyana	0.18	24.24	24.37	6	756	666	763	547	38	202
Haiti	6.41	70.07	52.90	222	2421	1643	9876	5241	2358	5031
Honduras	1.34	19.95	15.98	49	737	505	7318	3814	877	1067
Hungary	0.00	0.04	0.05	3	26	34	10 012	3252	0	227
Iceland	0.00	0.35	0.18	0	27	30	315	24	0	0
India	515.54	46.15	39.99	17 445	1562	1246	1181 411	833 320	133 332	741 655
Indonesia	36.05	16.15	16.15	1169	524	483	227 345	110 148	31 943	101 336
Iran (Islamic Republic of)	7.84	11.42	12.33	300	437	442	73 311	23 096	0	23 096
Iraq	26.84	97.75	98.05	814	2964	2121	30 096	10 075	4534	5138
Ireland	0.01	0.35	0.25	1	30	32	4436	1715	0	16
Israel	0.09	1.39	1.02	3	38	36	7051	587	0	11
Italy	0.05	0.08	0.03	13	23	29	59 603	19 063	0	0
Jamaica	0.25	9.34	8.35	9	324	284	2707	1265	139	670
Japan	1.48	1.16	0.51	36	28	34	127 293	42 708	0	2135
Jordan	0.52	9.60	7.59	21	396	282	6135	1324	119	277
Kazakhstan	3.35	22.16	25.47	116	768	880	15 521	6540	654	4970
Kenya	26.68	76.95	61.31	888	2561	1498	38 765	30 411	14 597	26 762
Kiribati	0.03	28.61	25.81	1	957	769	96	48	0	48
Kuwait	0.00	0.18	0.31	2	77	76	2919	48	0	48

Kyrgyzstan	1.43	27.67	25.18	51	996	905	5413	3451	518	2277
Lao People's Democratic Republic	2.59	46.51	46.57	80	1442	1078	6205	4290	2102	4118
Latvia	0.00	0.04	0.03	1	25	33	2258	721	29	295
Lebanon	0.26	6.46	6.35	11	271	257	4193	545	0	545
Lesotho	0.41	20.70	28.64	10	490	416	2049	1528	291	1452
Liberia	7.37	220.06	167.98	239	7125	4311	3793	1510	741	1510
Libyan Arab Jamahiriya	0.50	8.58	8.14	20	342	284	6294	1415	0	1415
Lithuania	0.00	0.13	0.21	1	29	40	3320	1098	0	1098
Luxembourg	0.00	0.95	0.70	0	36	42	480	84	0	1
Madagascar	22.85	125.98	101.37	735	4053	2530	19 110	13 479	9571	12 940
Malawi	20.66	160.26	114.06	674	5231	2962	14 846	12 060	2774	11 819
Malaysia	1.09	4.33	3.73	54	215	181	27 014	7976	80	717
Maldives	0.05	18.22	16.57	2	704	609	305	189	26	185
Mali	25.47	226.08	156.93	833	7391	4214	12 705	8620	4827	8534
Malta	0.00	0.10	0.05	0	24	29	407	23	0	0
Marshall Islands	0.02	27.41	23.54	1	942	751	60	19	0	19
Mauritania	2.55	88.60	79.16	79	2745	1832	3215	1898	1007	1632
Mauritius	0.01	1.22	1.39	2	127	133	1279	737	8	7
Mexico	5.26	5.09	5.43	238	230	209	108 555	24 720	3213	6921
Micronesia (Federated States of)	0.02	15.64	14.33	0	326	253	110	85	0	85
Monaco	0.00	1.35	0.80	0	37	42	32	0	0	0
Mongolia	0.58	22.62	23.94	21	830	811	2641	1133	578	1110
Morocco	4.87	16.15	16.63	170	562	512	31 605	13 904	5562	11 262
Mozambique	18.96	94.44	75.33	606	3016	1766	22 382	14 133	10 034	13 992
Myanmar	22.77	47.86	50.09	738	1553	1551	49 563	33 417	10 360	32 749
Namibia	0.21	10.48	21.26	6	281	263	2129	1346	161	982
Nauru	0.00	14.61	12.18	0	545	435	10	0	0	0
Nepal	15.78	59.41	49.86	522	1966	1345	28 809	23 853	3101	21 468
Netherlands	0.11	0.66	0.34	5	28	31	16 527	2995	0	0
New Zealand	0.01	0.22	0.15	1	28	31	4230	569	0	0
Nicaragua	0.99	18.30	16.38	37	678	519	5667	2452	784	1790
Niger	44.88	350.43	231.09	1457	11 377	6314	14 704	12 283	7492	12 160
Nigeria	199.75	144.74	118.83	6487	4701	2961	151 212	78 089	45 292	76 527
Niue	0.00	0.00	0.00	0	79	67	1	0	0	0
Norway	0.10	2.11	0.79	1	30	32	4766	1074	0	0
Oman	0.15	6.16	6.75	7	268	226	2785	791	182	648
Pakistan	67.83	43.67	44.53	2115	1361	1072	176 952	113 047	14 697	90 437

(Continued)

Annex *(Continued)*

	Estimated total deaths from diarrhoea, (thousand)	Estimated deaths from diarrhoea per 100 000 population	Age-standardized death rates from diarrhoea per 100 000	Estimated total DALYs attributable to diarrhoea (thousand)	Estimated DALYs attributable to diarrhoea per 100 000 population	Age-standardized DALYs attributable to diarrhoea per 100 000	Population total (thousand)	Total rural population (thousand)	Number rural with unimproved water supply (thousand)	Number rural without piped supply to home (thousand)
Palau	0.00	5.59	4.91	0	252	206	20	6	0	6
Panama	0.28	8.94	8.31	12	368	319	3398	907	154	190
Papua New Guinea	3.18	53.56	44.74	107	1804	1128	6576	5755	3857	5582
Paraguay	0.63	10.82	9.91	25	427	328	6237	2476	842	1609
Peru	2.89	10.72	9.99	118	437	379	28 836	8249	3218	5362
Philippines	17.53	21.16	21.86	552	666	528	90 348	31 649	4114	23 737
Poland	0.02	0.04	0.04	10	26	33	38 104	14 785	0	591
Portugal	0.04	0.39	0.23	3	27	32	10 677	4325	0	0
Qatar	0.01	0.79	1.25	1	93	98	1280	55	0	55
Republic of Korea	0.11	0.23	0.25	61	128	130	48 152	8917	1070	3210
Republic of Moldova	0.04	0.91	1.40	2	57	79	3633	2119	318	1843
Romania	0.10	0.47	0.77	9	40	57	21 361	9784	0	7240
Russian Federation	0.83	0.58	0.74	57	40	54	141 394	38 457	4231	23 074
Rwanda	18.93	209.17	147.53	633	6988	4117	9720	7942	3018	7863
Saint Kitts and Nevis	0.00	9.18	9.21	0	270	264	51	32	0	32
Saint Lucia	0.00	0.96	0.99	0	128	124	170	123	2	123
Saint Vincent and the Grenadines	0.00	4.12	4.36	0	212	199	109	57	0	57
Samoa	0.02	11.20	7.99	1	335	227	178	137	0	137
San Marino	0.00	0.00	0.00	0	22	29	31	1	0	1
Sao Tome and Principe	0.12	77.40	49.71	3	1756	1078	160	63	7	52
Saudi Arabia	2.30	9.98	8.80	92	400	305	25 200	4627	0	4627
Senegal	14.70	128.11	102.23	459	4003	2530	12 211	7046	3382	6200
Serbia and Montenegro	0.00	0.03	0.04	3	28	34	9839	4727	94	1749
Seychelles	0.00	1.95	2.22	0	114	118	83	36	0	36
Sierra Leone	18.24	338.46	236.97	594	11 015	6679	5559	3462	2563	3427
Singapore	0.02	0.45	0.46	3	61	73	4615	0	0	0
Slovakia	0.01	0.14	0.13	1	27	35	5399	2349	0	140
Slovenia	0.00	0.16	0.15	0	25	33	2015	1038	11	10
Solomon Islands	0.08	18.07	21.91	3	604	408	510	419	0	419
Somalia	18.79	236.27	205.98	584	7341	4517	8926	5668	5158	5668

South Africa	14.14	29.73	36.15	409	861	780	49 667	19 503	4290	13 262
Spain	0.39	0.90	0.37	11	26	31	44 486	10 187	0	0
Sri Lanka	0.94	4.93	5.49	27	143	153	20 060	17 038	2044	13 290
Sudan	20.17	55.80	48.56	665	1840	1228	41 347	23 371	11 219	20 099
Suriname	0.08	17.06	21.16	2	435	420	515	129	24	71
Swaziland	0.60	54.21	59.37	19	1707	1279	1167	877	342	693
Sweden	0.09	0.98	0.36	2	27	31	9204	1424	0	0
Switzerland	0.04	0.54	0.24	2	26	31	7541	2002	0	20
Syrian Arab Republic	1.83	9.93	9.36	71	385	284	21 226	9731	1557	2822
Tajikistan	5.42	83.84	55.98	189	2918	1944	6836	5030	1962	3772
Thailand	13.11	20.96	23.46	281	449	504	67 386	44 989	900	27 443
The former Yugoslav Republic of Macedonia	0.02	0.97	1.20	1	52	67	2041	675	6	108
Timor-Leste	0.28	27.69	16.77	11	1037	556	1098	798	296	710
Togo	5.37	88.50	75.12	174	2867	1789	6458	3743	2209	3706
Tonga	0.01	9.85	7.90	0	388	297	103	77	0	77
Trinidad and Tobago	0.03	2.29	2.68	2	137	147	1333	1157	81	300
Tunisia	0.95	9.46	10.99	31	313	334	10 169	3406	545	2078
Turkey	7.60	10.55	10.58	261	363	345	73 914	23 120	924	1849
Turkmenistan	2.83	59.42	52.97	96	2023	1774	5043	2592	0	2592
Tuvalu	0.00	21.88	19.21	0	738	583	9	5	0	0
Uganda	30.65	109.37	76.02	1034	3691	1920	31 656	27 555	9920	27 279
Ukraine	0.09	0.19	0.32	14	30	43	45 992	14 743	443	11 057
United Arab Emirates	0.05	1.23	1.70	4	107	118	4484	992	0	297
United Kingdom	1.53	2.55	1.03	19	32	33	61 230	6144	0	122
United Republic of Tanzania	36.89	98.36	92.20	1150	3066	2084	42 483	31 662	17 414	30 712
United States of America	4.58	1.54	0.86	97	33	34	311 666	56 985	3419	30 772
Uruguay	0.08	2.45	1.66	4	121	126	3349	257	0	21
Uzbekistan	9.83	37.50	30.79	349	1331	1096	27 191	17 202	3268	12 729
Vanuatu	0.02	8.46	7.91	1	334	236	233	176	37	118
Venezuela (Bolivarian Republic of)	1.78	6.76	7.69	72	274	244	28 120	1843	0	1843
Viet Nam	6.79	8.10	8.15	264	315	296	87 095	62 862	5029	57 204
Yemen	16.59	81.03	69.05	539	2633	1545	22 917	15 897	6836	13 194
Zambia	14.40	127.78	95.44	479	4247	2448	12 620	8158	4406	8076
Zimbabwe	7.60	58.33	63.77	231	1771	1407	12 462	7818	2189	7427

6

Defining the current situation – observing livelihoods

John Cameron

Around 1990, the analytical language used to describe rural people's lives in poorer countries and design policies to improve their lives shifted towards the livelihoods framework (Cameron, 1999; DFID, 1999). This framework aims to capture the totality of people's activities and the assets they use, and how they decide to change those activities when faced with, for instance, interventions improving their drinking-water situation. Though originally developed for poorer households, the framework is in principle applicable to any household – rural or urban, in poorer or richer economies (Bebbington, 1999; Carney, 1998). Using this framework, the impact of water interventions can be traced through a chain of linkages in which time and energy released by less illness, and reduced time spent on collecting and processing water, plus funds no longer spent on buying medical treatment, are used in additional activities that contribute to increased well-being (Dercon & Krishnan, 2000).

© 2011 World Health Organization (WHO). *Valuing Water, Valuing Livelihoods*. Edited by John Cameron, Paul Hunter, Paul Jagals and Katherine Pond. Published by IWA Publishing, London, UK.

USING THE LIVELIHOODS FRAMEWORK

Economic assessment of the impacts of rural water interventions is challenging. It requires judgements, implicit or explicit, on people's motivations and aspirations. Economists tend to assume that individuals are rational actors and will generally choose activities that offer economic gains. In the livelihoods framework, it is normal to assume that the household behaves like a rational individual and that decisions are benign in terms of protecting all its members' well-being. This assumption can be uncomfortable in terms of gender and age discrimination, but the framework can be sensitized to take on board such concerns. It is therefore argued here that using the livelihoods framework to assess changes resulting from small-scale drinking-water interventions is a useful option.

The livelihoods framework focuses on changing patterns of the natural, produced, human and social wealth that provide for those people's livelihoods and well-being (Blaikie, Cameron & Seddon, 2002; Whitehead, 2002). People with low incomes make changes in an attempt to cope with uncertainties stemming from nature and external human agencies. Often with minimal assistance from outsiders, they develop relationships which combine trust and discipline with flexibility in order to face the vagaries that the powers of nature and humankind can produce. Data collection methods such as participatory rural appraisal, and its successor participatory learning and action, can be seen as techniques for identifying these coping strategies (Chambers, 1994).

Table 6.1 indicates the types of benefits that the livelihoods framework suggests might be included in an economic assessment of a small-scale drinking-water intervention (see also Soussan, 1998).

There is a widespread perception that limited access to natural, produced and social wealth for many groups of people may mean extensive underemployment of their potential labour time. This would suggest a low value for their labour because the value of time freed up (in economic parlance, the opportunity cost) is close to zero in terms of productive work. Such a low value (in economic parlance, the shadow price) means that water interventions that increase the time available for other activities are unlikely to be economically attractive.

But people do move towards higher rates of earnings, and in that sense there is a healthy dissatisfaction with the generally low wage rates prevailing in more remote or in poorer economies (Ellis, 2000). Flexibility and fluidity of labour use reaches into all sectors. For instance, in rural economies, agricultural and non-agricultural activities often appear to be complementary rather than competitive in terms of their demands on labour time. A wide range of non-agricultural activities appear to be undertaken in agricultural slack periods alongside food-for-work public works (Benjamin, 1994).

Table 6.1 Possible livelihood benefits of providing small-scale drinking-water improvements for rural populations

Effect	Socioeconomic implications	Possible indicators
Increased water availability enhances natural wealth	Natural wealth available for greater use in a sustainable manner	Higher agricultural production in terms of crops, livestock and forest products
Experience gained with using produced wealth in the form of new drinking water technology	Incentive to acquire mechanical skills and new technology for other activities	Adoption of new technologies in other activities increasing productivity
Improved health of economically active individuals	More time and energy available for economic activity	Increased economic activity and additional time in value adding occupations
Less time spent caring for sick family members and fetching water	More time and energy available for productive, reproductive and social activities	Additional time in useful activities with direct and indirect gains to the household and society
Fewer infant and child deaths	Gain in net lifetime earnings	Average net lifetime earnings
Young people's school attendance improved	Gains from improved educational standards	Improved access to higher earning occupations
Nutritional gains, with improved absorption of nutrients or additional food production	More time and energy available for a range of activities	Dietary observation and anthropometric measures
Collective activity in planning and implementing improvements	Spin-offs to other local, collective projects	Evidence of consequent successful, local collective activities

It is to the credit of low-income people that they have maintained low levels of formal unemployment by showing a capacity for creating income opportunities with little or no non-human resource investment. Recognizing the dynamism and mobility of people in the broadly defined "non-formal" economy in both rural and urban areas has been a feature of the development debate for more than 20 years. In terms of activities, relatively low-productivity trade and personal services tend to dominate livelihoods diversification, followed by micro-scale construction and low-technology transport. In terms of the spatial dimension, much of the pressure of livelihoods improvement for low-income people, who have proven difficult to reach in terms of the Millennium Development Goals, is being expressed through internal migration of individuals or whole households, or their attempted movement towards neighbouring or more distant economies.

Thus in order to assess the opportunity cost of and thereby put a shadow price on labour in the catchment area of a water intervention in a poor rural location, the following types of information on the general status of the local economy will be useful:

- What is the proportion of households receiving incomes from family members working outside the intervention catchment area?
- What is the proportion of households receiving government grants?
- What is the proportion of households deriving a cash income from agricultural activities?
- What is the proportion of households deriving a cash income from activities inside the intervention catchment area?
- What is the proportion of households who invested in house improvements or vehicle purchase, or tools and equipment in the last year?
- What is the mean annual total cost to a household of sending a child to school and what proportion of school-age children are in school?
- What is the mean annual total cost to the government of providing a school place (as an indicator of schooling quality)?

If the economy of an intervention catchment area has diversified activities, is well connected to wider labour markets, and has strong investment in technology/produced wealth and children's education, it will merit a shadow price close to GNP per capita rather than either zero (no economic value of time freed by an intervention) or the going wage rate in the area. Thus collecting data on livelihoods does not only require an understanding of the local economy, but also how that economy is connected to wider national and even international economies.

REDUCING INFORMATION ERROR IN A COMPLEX LIVELIHOODS CONTEXT

Collecting information on livelihoods is very demanding, as the concept is concerned with a holistic, dynamic, integrated image of a household in its local and wider context (Laws, Harper & Marcus, 2003; Mikkelsen, 1995; Thomas & Mohan, 2007).

Livelihoods questionnaires can run to dozens of pages and hundreds of variables. Information collection on livelihoods also requires an awareness of the social relations in which households are embedded, as these may be better investigated using more collective, qualitative, pictorial techniques in collective or participatory interviews (Cameron, 2006). Many of the challenges involved in collecting accurate information, described in Chapter 5, apply to livelihoods investigations. But there are wider possibilities of error in livelihoods observations as a result of greater dependency on human testimony for numerous variables. Both conceptual and practical challenges in generating acceptably accurate information on livelihoods are the focus of this chapter.

Livelihoods investigations frequently combine information gained through physical measurements with information directly gathered from participants or beneficiaries in the form of testimonies from individual and household structured or semi-structured questionnaires plus records of group meetings. Testimonies yield information about otherwise inaccessible past experiences and perceptions of possible futures. The gathering of testimony information can produce richer information and encourages ownership of the activities by the people giving the testimonies, but only if collected in a way that respects them (Cooke & Kothari, 2001).

Testimonies are bound to be influenced by the values and interests of those making them. In addition, even if values and interests are minimal influences, the datasets will be only as accurate as people's memories allow. There are clear limits to how far testimonial evidence can replace physical measurements. Therefore, the aim should be to combine data collection techniques in imaginative ways that are cost-effective in terms of acceptable accuracy given the scale of the intervention being planned.

The challenge of any information collection activity in practice is to combine a concern for all the different potential sources of error with an overview of the whole information gathering process. In practice, any combination of more qualitative and more quantitative methods can be applied well or badly in terms of how all potential forms of error are handled. Quantitative methods may generally have concerns with measurement inaccuracies, while qualitative methods may have concerns with representativeness, but neither can claim immunity from any of the various forms of error to which all data-collecting and information-reporting exercises are prone.

In simple terms, the methods can be seen as complementary, with the quantitative aspect seeking to measure typical or, average values of variables, while the qualitative aspect seeks to reveal the whole range of human experiences involved. These experiences both help explain and create further complications in understanding the average values.

Sources of error in livelihoods investigations can be usefully subdivided into a number of types which are associated with stages in the information process. These appear to come at the information gatherer in a chronological order:

- formulating the deep issues and establishing the baseline – the risk of mis-specifying concepts and causalities in the pre-intervention situation;
- identifying objectively measurable variables – the risk of perversely moving indicators or proxies in relation to the underlying concepts, e.g. decrease in reported incidence of diarrhoea episodes as an indicator of improved livelihoods if the time and money is spent in alcohol consumption;
- designing the observation instrument – the risk of inaccurate measurements;
- identifying the relevant population of cases – the risk of mis-specifying who or what is being represented;
- deciding which cases to investigate – the risk of bias and the handling of formal sampling error;
- negotiating contracts for detailed specific funding or resources – the risk of under-resourcing and the influence of vested interests;
- employing, training and supervising the people who collect the data – the risk of poor quality interviewing or of cheating;
- processing data – the risk of transcribing and data-input error;
- analysing data to produce information – the risks of inadequate and inappropriate tests and interpretation error;
- communicating information – the risk of a policy-maker or decision-maker misunderstanding the implications of the results.

Each of these stages has its own potential as a source of error. The design of each stage also has a capacity for increasing or reducing error at later stages in the information gathering process. Therefore, good information gathering on the livelihoods implications of drinking-water interventions demands attention to the process as a whole, as well as to the individual parts.

The most telling example of the interconnection between later errors and earlier decisions is the division of labour in which an external consultant designs a survey up to the point of fieldwork, and then leaves, only to return for analysis and reporting. Lack of concern for the intervening interviewing and processing errors at the earlier stages of design of the measurement instrument and sampling frame can produce inaccurate results (rubbish in, rubbish out), which remain undetected until someone else attempts to use the results practically for policy or research.

In conclusion, in even the most perfectly designed exercise to collect information, if something can go wrong, it will. Each problem anticipated, contractually agreed, and budgeted is a problem virtually solved. Each problem left to be "sorted out later" is a hostage to ill-fortune with a potentially high ransom payment in cash and good-will.

The following sections go systematically through the livelihoods information collection process. The process is described primarily in terms of a questionnaire survey, but the principles are also applicable to direct physical engineering and health measurements, as well as more ethnographic information collection.

RISKS IN FORMULATING THE DEEP ISSUES AND ESTABLISHING THE BASELINE

Livelihoods surveys and studies seek to understand individual and household decision-making in order to establish the significance of interventions. Formulating the deep issues and establishing the baseline pose risks of incorrectly specifying concepts, causalities and the pre-intervention situation.

Building a picture over time can be attempted using a variety of forms of comparative baseline data, all of which may present problems, for example:

- a purposive baseline study – when used later in the implementation process may involve problems of hardware and software compatibility plus loss of details on sampling procedures, including location of cases if administrative boundaries have changed;
- secondary reports on the intervention area prior to the intervention, presenting processed data in extensive tabulations – may involve problems of inadequate reporting on the sampling frame and an inadequate basis for valuation if price inflation has occurred;
- unprocessed primary data on the intervention area prior to the intervention – may involve problems of illegibility and interpretation of how questions were actually asked, and coding into the format of the current questionnaire can be demanding in resources;
- secondary reports with few data tabulations – results of current information collection will have to be interpreted in terms of the conceptual framework of the original reports;
- questions in a later questionnaire during or after implementation asking respondents to recall the baseline situation – problems of accurate recall and rationalization of past experience in the light of current attitudes.

Therefore, whichever form of baseline data is used, there are bound to be conceptualization and accuracy questions about what was really the position in the past. Flexibility and a hard-headed, robust approach are needed to decide what is going to be truly comparable and what can be safely reported about changes.

In creating a baseline, there will also be an element that theoretically seeks to link the intervention to impact on livelihoods. Drinking-water interventions connect to livelihoods through decisions on how to use released time and energy. This involves assumptions about people's motivations, attitudes to risk and aspirations. These assumptions need to be made explicit at this stage.

IDENTIFYING OBJECTIVELY MEASURABLE VARIABLES

Relevant variables are relatively easy to identify. But identifying reliable observable indicators for these variables can cause problems. For instance, demographic characteristics of households are needed to determine per capita costs and benefits, but identifying who is, or is not, in a particular household, for the purposes of attributing shares of income or consumption, requires care.

Assessing household health and educational status requires multiple indicators. Information on the use of health and education services needs to be combined with outcome indicators (e.g. cures and credentials) to assess the effect on well-being. It is difficult to agree on easily measurable indicators to reflect the impact of poor quality drinking-water on human well-being (see the later discussion on the challenges of using DALYs).

The World Bank Living Standards Measurement Study methodology and its numerous supporting documents provide useful guidelines for the selection of variables and indicators, including some health indicators. But these guidelines should be implemented in the context of national or local circumstances.

It is vital to think imaginatively (and collectively) about all indicators, in order to identify possible circumstances in which one or more of the indicators may move in a positive direction while the underlying organizational capability or livelihood trajectory moves negatively into greater vulnerability. The possibility of an indicator concealing a move into greater vulnerability should be considered from the point of view of whole organizations, households, and individual members of both.

DESIGNING THE MEASURING INSTRUMENT – THE RISK OF MEASUREMENT INACCURACY

A judgement is needed on whether the people to be interviewed are likely to be self-confident, articulate and well-informed, and therefore capable of responding to an open interview permitting wide-ranging responses, rather than a more closed, formal questionnaire. Processing considerations are also important in deciding the degree of closure of the questionnaire. Pilot interviews are always essential with the whole range of likely interviewees, plus coding frames and trial tabulations.

Every question in a questionnaire needs to be thought about in terms of its intentions (it is worth for this purpose distinguishing between "data" as an observation with no clear use and "information" as data converted into a useful piece of evidence). For every data item, it is worth considering:

- Are data providing variables for direct reporting as information?
- Will data be combined with other data (or extended beyond the response period investigated) to give a composite or longer-term variable for reporting as information?
- Will data assist recall on a subsequent question which will produce information to be reported?
- Are data needed to check the consistency of other information?
- Is a question being asked to ease the conduct of the interview, when the data produced by the answer are very unlikely to be reported as information?

Such considerations will help the analyst set priorities when, inevitably, the draft questionnaire gets too long. They will also help in deciding on the order of questions. A lack of focus on why particular data items need to be collected produces a sprawling questionnaire, difficult to apply, difficult to answer, and difficult to process. A useful way of thinking about a questionnaire (or the conduct of a participatory data collection exercise) is to think about it as being designed by a group of four people: the interviewee (respondent); the interviewer (enumerator); the processor; and the analyst. It is vital to get the priorities right:

- The interviewee (respondent) is the most vital person and has the top priority. Language, units of measurement, order of questions, and length of questionnaire must all be designed to make literal and cultural sense to the respondent. It is important that this applies to the whole range of respondents. For example, rural livelihoods questionnaires tend to focus on middle-income households with income primarily derived from their own land, which means that the questionnaire is much less relevant for households with non-cultivation incomes. The use and positioning of more qualitative questions (even if it is unlikely that the responses will be systematically processed) can help the respondent feel less interrogated. Avoiding asking leading questions will prevent the respondent from lapsing into easy affirmatives. Similarly, the form and positioning of more sensitive questions is important.

- The interviewer (enumerator) is the second priority. Is the person administering the questionnaire an interviewer expected to look around and to make a judgement on acceptable tolerances in responses? Or is the person administering the questionnaire an enumerator expected to ask the questions doggedly and record the answers mechanically? The questionnaire needs to be appropriate to the skills and experience of the interviewers, who should feel comfortable at the beginning and at the end of the interview so they feel motivated to go on to the next interview, and able to return if follow-up is needed. A poorly designed questionnaire can be satisfactorily administered by an over-qualified interviewer. That is, however, no more an excuse for bad design than the fact that a badly designed car can be safely driven by a highly qualified driver.
- The processor needs all the assistance that can come from a well laid-out, pre-classified questionnaire. Processing is hard work and a clear layout with pre-coded columns, which the eye can easily follow, is vital to efficient, accurate processing. If a questionnaire is to be translated, it is vital that the translator and printer are clear that layout is an important consideration in the final questionnaire. There are simple rules of clarity, such as avoiding ambiguity about meanings of marks and non-marks on the questionnaire. For example, there needs to be agreement about the way to process crossing out and ticking in, dashes and negatives, and zeros or refusals to answer or non-responses.
- The analyst is vital to identifying the data items to be collected, but having done the task, the analyst should be excluded from the design of the observation instrument. If the analyst is also the designer, then the analyst role should be mentally set to one side. The analyst may have clear causalities and a sense of the logic of the final information in mind, but there is no reason why these should determine the shape of the questionnaire. If the questionnaire is to be translated, then the analyst has a legitimate interest in ensuring that the questionnaire is independently back-translated to check that the specified data are actually being collected.

The questionnaire needs to be appropriate for an interview that will often be at least a household interview and that may include other people. The interview will be a social event, perhaps involving up to twenty people – and possibly including people in a power relationship with the target household. The interview may also be interrupted by lively children or noisy livestock.

RISKS IN IDENTIFYING THE RELEVANT POPULATION

In the identification of the relevant population, there is a risk that those whose experiences are represented are incorrectly specified. Generally, in collecting information on a drinking-water intervention, the population is defined as all people resident in a specified catchment area. It is usually possible to identify the households in an area, although the place of residence of nomadic households will be ambiguous. The wide variety of potential forms of migration from "settled" households, however, can also create ambiguities about the people who can be considered the target population to be served by the intervention.

There may be a wish to disaggregate the households or individuals into strata – with varying degrees of confidence in the statistical validity of the subdivision. The precise populations for the purposes of stratification are often not known, especially if wealth statistics are likely to be inaccurate because of legislation, landrights or taxation. Therefore, the basic principles of stratifying the population should be identified before the sampling method is decided and fully documented.

If there is a baseline survey, then there may be confusion over precisely what is being followed up. There will be a temptation to follow up the previous interviewees as a cohort population. A cohort approach has a possibility of greater qualitative insights and cross-checking against original data. But it has a clear bias in not representing the current situation of a wider population, because it concentrates on the "survivors" in a particular location.

Household heads who are survivors are likely to have been younger, healthier, more economically secure, or less adventurous than those who died or migrated. This bias can be only partially offset by selecting descendants of the previous household heads who still live in the original locality. In many locations, this approach is not only biased against male descendants who have migrated, but also strengthens the gender bias, as daughters almost invariably change location on marriage.

Before gathering information on socioeconomic development, careful thought has to be given to precisely whose lives are being tracked.

DECIDING WHICH CASES TO INVESTIGATE – THE RISK OF BIAS AND SAMPLING ERROR

Arguably, formal sampling error receives too much attention as a form of error in manuals and texts on surveys. The certainty of both avoiding bias through random sampling and making conditional statements on representativeness are very attractive islands of science in an ocean of judgement about other forms of error.

The statistical theory of random sampling also offers help in deciding how many cases to investigate. An approximate sample size can be calculated from information about the variability of some major variables, along with views on acceptable levels of inaccuracy. But judgement is involved and, in the final instance, the degree of error that is acceptable will depend upon how uncomfortable the reported information is to the reader or decision-maker. Nevertheless, taking as random a sample as feasible is a useful first line of defence against accusations of bias, though not against accusations of non-representativeness.

Deciding how far to cluster, as opposed to stratify, cases is a product of judgements on how to best manage and budget fieldwork. The rule of thumb of doubling the number of cases to compensate for the effect of clustering can produce a complex trade-off depending on the costs of moving interviewers or enumerators and their supervisors around. It is important for analysis that a clear demarcation is made between cases where randomness has been applied wherever feasible, and where clustering or more purposive sampling has been used for other reasons.

NEGOTIATING CONTRACTS – THE RISK OF UNDER-RESOURCING AND INFLUENCE OF VESTED INTERESTS

Any exercise to collect information is likely to involve relationships between various funding agencies and implementing partners. Agencies and partners will have different needs and capacities. In best practice, every stakeholder would be involved at the conception of the exercise, fully participate in every stage of the information process, and own (in every sense) the results.

Funding agencies will want to know whether their original budget is now likely to be exceeded and why. International and national agencies will try to assess each other's institutional rewards from the intervention, well aware that such rewards may be vastly different even though inputs are similar. It is vital to sort out contractual and management relationships before recruiting interviewers or enumerators and going into the field. Employment in many countries is rightly not a casual affair and all employees have a right to know who is responsible for paying them and, when necessary, disciplining them. Information collection in remote rural areas needs logistical support and can involve health risks, so responsibilities for insurance need to be clear.

Processing, analysis and reporting need adequate resources (software and hardware) and must be fully budgeted. Maintenance and eventual ownership of any hardware needs to be clear. Processing especially is frequently an under-resourced activity. Also dissemination activities need to be planned and budgeted, especially if international travel is involved. Knowing that the work eventually will receive national or international attention can be an important motivator when logistics go wrong in the field.

A careful choice of partners is vital. Fortunately there are now many organizations used both to commissioning and to implementing information collection. These include government agencies, non-governmental organizations engaged in development work, and private sector companies that carry out market research. Unfortunately, there are also individuals and organizations who regard conducting surveys as a simple source of revenue. These unscrupulous agents recruit some underpaid school-leavers, send them out to a few intervention catchment areas, input the data without accuracy checks, output undigested tables of raw data, and collect the fee with maximum contingency payments. The results will meet strict contractual terms; what will be missing is good quality information.

There is clearly a case for a full partnership in which everyone is a stakeholder. Working together on information collection can provide mutual gains in terms of national or international status, experience that can be used in advertising, skills training (surveys are an essential aspect of good management information systems), and general organizational or managerial capacity building.

RISKS IN FIELD MANAGEMENT

In field management there may be risks of undermining the sampling frame and losing contextual information. Recruitment and training of trustworthy interviewers is an initial consideration and ensuring their effective support is vital. Flexibility in the field may be required in the following areas:

- Selection of cases in the intervention catchment area – ensuring that the population listing is complete and that there is a random selection of households for interviews should involve the whole team. Involving key informants from the intervention catchment area can dispel fears that selection was based on unspoken preferences or prejudices (though random selection can bring its own problems in cultures where chance is viewed with suspicion).
- Field substitution of households or individuals who refuse to be interviewed or are unavailable for interview – interviewers or enumerators need clear substitution rules to ensure representation from groups of "hard to find" or "hard to interview" households.

- Using the eyes and ears of interviewers or enumerators for more than recording responses – observant interviewers or enumerators can record qualitative information on the questionnaire or at a participatory data collection meeting about causalities and unresolved inconsistencies. For example, assets may be seen that are inconsistent with claimed poverty, or multi-activity households may declare that they have produced large amounts of crops, but these crops may not be visible in fields or in storage (such inconsistencies will presumably stem from some form of fiscal fraud, such as money-laundering). In addition, consistency calculations on aggregate crops and cash flows can be made in the field by aggregating data from various parts of the questionnaire.
- Interviewers or enumerators may find it useful to modify the order of asking questions to encourage responses. For example, if people of different genders or ages are being interviewed it may be useful to start with the questions that are most relevant to their specific activities.
- All interviewers and enumerators can be encouraged to keep a full field diary to record impressions of contexts and methodology, and time should be allowed for debriefing on field diaries.

The analyst can gain much from observing power relationships, and emotional tensions in the spaces or silences, which will not be recorded in questionnaires. Even with limited local language ability, the analyst in the field can observe how spatial organization of meetings and body language can act to exclude participation, and hence bias data against more vulnerable people's experiences.

RISKS IN EMPLOYING, TRAINING AND SUPERVISING INTERVIEWERS OR ENUMERATORS

The manner in which interviewers or enumerators are found and recruited is important to their subsequent commitment. They are more likely to work well if they know they have been selected on merit, rather than through nepotism. A recruitment policy that explicitly offers equal opportunities, plus affirmative action, should ensure the representation of women and minority groups.

If the questionnaire resembles a semi-structured interview schedule, with a relatively open, qualitative style, it will require interviewers with professional knowledge and mature personal skills, and they should have responsibility for analysis and reporting. But any good interviewer or enumerator will have a personal style that is attentive not egoistical, and authoritative not authoritarian. Social skills are as important as formal qualifications. In general, it is better to recruit fewer people for longer periods of time, and involve them in quality assurance and processing.

Decent rates of pay, meeting all living costs in the field, and contractual security all act as incentives to a professional approach. While local knowledge is important and local language is essential, it is important that the first loyalty of the interviewers and enumerators is to the information collection, not their peers in the locality. Involvement in pilot-testing and modifying the questionnaire, and understanding the sampling method (needed for field substitutions) are not only vital aspects of induction training but also encourage a sense of ownership.

A clear management style needs to be set from the outset. Given that interviewing is individual and dependent upon the vagaries of human nature, interviewers and enumerators cannot be supervised at all times, and they will have to exercise judgement. A total quality management approach is therefore indicated, emphasizing responsibility to a stable team, a no-blame culture, respondent-centred quality, and well-documented innovation.

This approach can be contrasted with a total quantity administration style in which interviewers or enumerators are set individual targets for numbers of interviews and rewarded for returning the set number of questionnaires to a central administrator, with incentives for early completion.

Teams of five or six people can be accommodated in most intervention catchment areas. Teams of that size suit both quantitative and qualitative work, and can accept temporary or permanent losses, and induct replacements without disruption. The presence of the analyst can be useful in terms of morale and technical advice, but care needs to be taken not to undermine day-to-day management. Management needs to be aware of interviewers or enumerators becoming tired or jaded. This may be revealed in body language, tone of voice or irritability. The possibility of returning home after being away in demanding circumstances needs to be built into the phasing of fieldwork – not to do this may discriminate against employing women with young children.

It may be useful to identify five models of dysfunctional interviewing, in order to help correct habits that may understandably be acquired during periods of continuous interviewing – habits not appropriate to the collection of good quality information:

- task completion – competing over how many interviews have been completed in the day;
- the quiz – helping respondents to get the "right" answer by asking questions in a leading manner;
- therapeutic counselling – giving advice on medical, financial, career and personal or political relationships;

- the mantra – asking questions without looking at the interviewee and without varying intonation;
- dysfunctional joking – gossiping to other team members or elite insiders from the locality, and making some respondents feel inferior and marginalized.

Conduct in the field must follow the social norms of the people being interviewed. There will be a need to frequently explain the objectives of the information collection and why this group of strangers are in the intervention catchment area. It cannot be assumed that an intervention catchment area is networked for flows of accurate information. Rumours travel faster than facts, as they are usually much more exciting. Hinting at policy benefits from co-operation is in nobody's interest, though it will be tempting, especially for any government officials or nongovernmental organization staff in the team.

PROCESSING DATA

Every act of transcription introduces possibilities of error. Questionnaire design can play an important role in minimizing this form of error by cutting out the need to use intermediate coding sheets before data entry into a computer.

Coding on the questionnaires in a designated column with pre-set coding boxes, preferably by the interviewer or enumerator during or immediately after the interview, is highly desirable – provided the complexity of the coding system does not disrupt the interview process. If this direct approach is adopted, then interviewers and enumerators will need training in distinguishing zero values of variables from non-responses in coding.

The format for data entry needs to be simple, which paradoxically means that it may be very lengthy, with every mark on the questionnaire having its own field. The act of manual data entry will be replaced eventually by machine scanning, but for the moment it is doomed to be a repetitive task, with the human being involved acting as a mere operative. Processing works best when the data go from eye to hand with minimum thought.

The Fordist (with strict division of labour) nature of the process leads to attitudes on both sides of the employment relationship. Issues arise such as:

- pay incentives for quantity and quality of work;
- flexible working hours to ensure that employees work when they are most willing;
- adequate working conditions in terms of comfortable furniture and temperature, good lighting, and minimum distraction in terms of noise;

- health and safety issues, especially screen exposure and repetitive strain injuries.

A logical extension of this Fordist model will raise the question of whether the processing task can be subcontracted out to a private sector or public sector agency, on a commercial basis. This raises ethical and quality control issues.

The ethical question is whether the contracting out seeks to reduce costs through poor employment terms and conditions – employment terms and conditions that would be unacceptable if the workers were directly employed by the agencies conducting the information exercise.

Quality control is important whether processing is carried out in-house or by sub-contractors. Quality control can range from complete double-entry, through random checks, to reliance on software to report extreme values. The more Fordist and the more sub-contracted the processing work, the more extensive, rigorous and costly should be the quality control procedure.

For most state-of-the-art computer software, from databases through spreadsheets to specialist sector-specific analysis packages, there is no clear frontier between processing and analysis. In practice, procedures which use separate software packages for data entry and analysis are likely to have translation problems, especially when electronic mail is used to transfer data and tabulations.

Because generating tables and statistics does not require specialist programming skills, analysts may be tempted to intervene in the processing and transform raw data into digestible variables at the moment of data entry. On balance, this is probably not a good idea, except perhaps for smaller exercises to collect information, where data entry and producing tables are being performed by a small team working closely together.

Processing more qualitative data, gathered through ethnographic and participatory methods, involves judgements on significance that can only really be made by the analyst – one of the major reasons for restricting such data collection to a few case-studies. There is software for analysing transcripts of qualitative data e.g ATLAS/ti and NVIVO, but it is important to remember that judgements by the analyst are still crucial and should be made explicit.

ANALYSING DATA TO PRODUCE INFORMATION

Many surveys and studies attempt to trace movements across time (and, to a degree, across space). A point-to-point mapping is possible if baseline information exists. The baseline may be in computer data files or hard copies of questionnaires or tabulations. In any form, the existence of a baseline can be unduly reassuring because it appears to reduce analysis to hypothesis testing of whether apparent differences can escape the grip of sampling error. But reality is never that simple. Original data files are likely to prove difficult to convert into exactly comparable form and the precise specifications of the data underpinning hard copy tables may be difficult to identify. An element of judgement, in addition to a calculation of sampling error, is bound to be involved.

Analysing any dataset to produce information demands judgements to be exercised, though arguably more for qualitative than quantitative data. Judgements are more likely to be accepted if made collectively by involving users, and more ethical if they are made explicit. Sensitivity tests reveal the judgements that have been and are vital in social cost-benefit analysis.

COMMUNICATING INFORMATION

Communicating the results to the eventual target groups – the people we would like to inform – presents a challenge. In small projects, information flows will be primarily aimed at the intervention management and the people directly affected by the intervention: in larger interventions, many more stakeholders are likely to be interested in the data generated, for example donors, research institutes and policy-makers.

There are two temptations in communicating: the first is to give the audience what they expect; and the second is to give the audience something that will surprise them. To recognize both temptations, and know which is which, is an important basis for good communication. Agencies commission studies or surveys for both confirmation and new insights. It is important to know what the information commissioners are expecting. It is also important to remember what the data providers (and collectors) understood about the objectives of the information collection when they responded.

The following comments should be borne in mind:

- It is never too early to communicate – this especially applies if the news is unwelcome.
- When communicating the conclusions, the related technical qualifications that arose in the analysis tend to get lost – decision-makers prefer to think that information removes all risks and makes the decision for them.
- People are exposed to a great deal of professional communication in a variety of forms, including audiovisual presentations – good communication needs imagination and adequate resourcing.

There are important ethical aspects to communication, concerning rights to knowledge and intellectual property rights. Organizational and livelihoods surveys and case-studies can be potentially dangerous to powerful interests. Information gatherers have a responsibility, as a point of principle, to negotiate the widest dissemination results, notably to the people who gave the data.

Lastly, it is vital that any results of an economic appraisal are presented along with an explicit disclosure of all the assumptions made, with sensitivity tests performed on a range of possible scenarios. Economic appraisal is not a precise calculation. It should feed into the decision-making process, not make the decision by claiming technical closure.

CONCLUSION

Following the guidelines in this chapter will facilitate estimating, with reasonable accuracy, the livelihood characteristics of a target population in a drinking-water intervention catchment area. These characteristics will cover the portfolios of assets which the target population possess, the way in which they convert these assets into activities, and the time and energy spent in water collection. Insights will also be gained into the division of activities between women and men, and between generations. Collective participatory methods can reveal the forms of social relationships and the social wealth in the target population, and how collective decisions are made.

The livelihoods framework can help identify the variables which are significant in estimating the incremental benefits attributable to the drinking-water intervention. Indicators of physical processes precede attributing values or prices.

REFERENCES

Bebbington A (1999). Capitals and capabilities: a framework for analyzing peasant viability, rural livelihoods and poverty. *World Development* 27: 1021–1044.

Benjamin C (1994). The growing importance of diversification activities for French farm households. *Journal of Rural Studies* 10: 331–342.

Blaikie PM, Cameron J, Seddon JD (2002). Understanding 20 years of change in West-Central Nepal: continuity and change in lives and ideas. *World Development* 30: 1255–1269.

Cameron J (1996). The challenge of combining quantitative and qualitative methods in labour force and livelihoods analysis: a case-study of Bangladesh. *Journal of International Development* 8: 625–653.

Cameron J (1999). Trivial pursuits? Reconciling sustainable rural development and the global economic institutions. London, IIED: 42.

Cameron J (2006). A participatory approach to evaluation of the impact of NGOs on development in Nepal. *Development in Practice* 16: 91–96.

Carney D, ed. (1998). *Sustainable rural livelihoods: what contribution can we make?* London, Department for International Development.

Chambers R (1994). Participatory rural appraisal (PRA): challenges, potentials and paradigm. *World Development*, 22: 1347–1454.

Cooke B, Kothari U, eds. (2001). *Participation: the new tyranny?* New York and London, Zed Press.

Dercon S, Krishnan P (2000). In sickness and in health: risk sharing within households in rural Ethiopia. *Journal of Political Economy*, 108: 688–727.

DFID (1999) *Sustainable livelihoods guidance sheets.* London, Department for International Development.

Ellis F (2000). The determinants of rural livelihood diversification in developing countries. *Journal of Agricultural Economics*, 51: 289–302.

Laws S, Harper C, Marcus R (2003). *Research for development – a practical guide.* London, Sage.

Mikkelsen B (1995). *Methods for development work and research: a guide for practitioners.* London, Sage.

Soussan J (1998). Water/irrigation and sustainable rural livelihoods. In: Carney D ed. *Sustainable rural livelihoods: what contribution can we make?* London, Department for International Development (DfID): 181–195.

Thomas A, Mohan G (2007). *Research skills for policy and development – how to find out fast.* London, Sage.

Whitehead A (2002). Tracking livelihood change: theoretical, methodological and empirical perspectives from north-east Ghana. *Journal of South-African Studies*, 28: 575–598.

7

Interventions for water provision

Stephen Pedley, Katherine Pond and Eadaoin Joyce

This chapter provides a summary of the interventions that are available to improve water provision to communities using small-scale systems. This chapter is intended for policy analysts who are new to the area of water supply management and who want to learn more about the range of issues that they may need to take into consideration when making policy decisions.

The chapter covers three topics: water interventions; education and community training; and policy and planning. It offers only a brief summary of these topics – a comprehensive review of current knowledge and practice is beyond the capacity of this book – but we have endeavoured to include easily accessible references that provide further details.

In contrast to the small amount of technical detail that we have included, the scope of this chapter is made deliberately broad in order to include interventions that cover the range of challenges that face both developed and developing

© 2011 World Health Organization (WHO). *Valuing Water, Valuing Livelihoods*. Edited by John Cameron, Paul Hunter, Paul Jagals and Katherine Pond. Published by IWA Publishing, London, UK.

countries. As a consequence, when reading the chapter it is important to consider the type of intervention in the context of the physical and institutional environment and of available resources, and to select interventions that are appropriate to the challenge that has to be met. For further support and information, see: http://www.who.int/water_sanitation_health/dwq/small community/en/index.html.

WATER INTERVENTIONS

Water interventions can take place at many points along the supply chain from source to consumer. For example, some communities may take water directly from a source, such as a river, and transport it home for storage and consumption. By incrementally introducing interventions such as source protection, mechanical abstraction, storage, treatment and distribution, the safety of the water can be improved and the level of access increased.

No matter what structure the water supply system takes, the priorities for improvement should be determined before any interventions are implemented. For this task, the most appropriate tool is the water safety plan (WSP), an approach launched in the third edition of the WHO *Guidelines for drinking-water quality* (WHO, 2004):

> *"The most effective means of consistently ensuring the safety of a drinking-water supply is through the use of a comprehensive risk assessment and risk management approach that encompasses all steps in water supply from catchment to consumer. In the Guidelines, such approaches are termed water safety plans (WSPs)".*

Following the first publication of WSPs, several reports have been published that provide guidance on their use and case studies of their application (Godfrey & Howard, 2005). WSPs provide a cyclical process for the continuous improvement of a water supply system irrespective of its size or complexity. The plan allows for appropriate interventions to be made to control potential sources of contamination. It introduces validation of control measures and a system for monitoring the control measures, and it allows for timely corrective actions to be made to protect the health of the consumers. The concept and principles of WSPs should be used to inform the selection of water interventions.

SOURCES OF WATER

The discussion below draws heavily on the work of (Carlevaro & Gonzalez, 2011). Natural sources of water are of three main types: rainwater (collected from roof run-off and ground surfaces); surface water (for example, streams, rivers, lakes, impoundments and reservoirs); and groundwater. Of the three, groundwater is often assumed to be of good quality as a result of the protection it receives from the overlying soil. This assumption is, however, not always correct. Contamination may enter the groundwater at vulnerable points, such as shallow, fractured soils or abandoned wells and boreholes. A comprehensive review of groundwater and groundwater vulnerability has been published by Schmoll et al. (2006). In contrast to groundwater, rainwater and surface water are considered to be more susceptible to contamination.

The selection of the water source will depend on a number of factors: its yield, reliability, quality and distance from the community that it must serve; whether it can be collected by gravity or if pumping is required; its vulnerability to natural hazards, such as flooding and freezing; and its accessibility. More importantly, the final choice of the source will depend especially on the quantity and quality of the water, the costs of development and operation, and the funds available.

In regions where treatment of the water source is impractical, it is imperative that the selection of the source takes into consideration, among other things, the safety of the water and the opportunities for protecting its quality against contamination. Accessibility is also an important consideration if the water is to be collected directly from the source, rather than conveyed through a distribution system to the user. In this respect, the WHO/UNICEF Joint Monitoring Programme on Water Supply and Sanitation recommends that a drinking-water source should be less than one kilometre away from its place of use and that it should be possible to reliably obtain at least 20 litres per member of a household per day. The process of selecting the water source should take account of the particular needs of low-income people, because they are at greatest risk of infectious diarrhoeal disease from inadequate water supply (Wilkinson, 1998; Eisenberg et al., 2001; Payment & Hunter, 2001; Howard, 2002a).

Finally, any improvement to the current water source must take into account the rationale behind the existing use of water sources (some sources are more reliable, convenient or simply taste better). If an "improvement" results in poor performance on any one of these aspects, people may return to their traditional source (Hanson et al., 2003).

The following sources of water will be reviewed briefly: rainwater; groundwater; surface water; and a water resource jointly developed by a number of communities in a particular region.

Rainwater harvesting

Rainwater harvesting is an ancient technology that has a proven track-record of providing water next to the house for domestic use and, on a larger scale, for economic use by increasing the productivity of arable lands and watering livestock

(Deepesh et al., 2004; Smet, 2005). It has also been used successfully to supplement alternative water supplies in small communities and institutions. In much of the developed world, rainwater harvesting declined in popularity after the introduction of the large-scale piped water systems. However, the present drive towards the use of sustainable environmental systems and the protection of declining freshwater reserves has revived interest in the use of rainwater for some domestic purposes. In a recent review of the potential benefits of rainwater harvesting in the United Kingdom, Rachwal & Holt (2008) point to many examples of the successful use of rainwater in Australia, Germany and the United States.

Smet (2005) has published a fact sheet for rainwater harvesting that presents an overview of systems, component technology, planning and management, and the potential effects of the technology. For a more thorough review of the subject, the reader is referred to one of the many publications dedicated to rainwater harvesting (for example, EnHEALTH, 2004; Pacey & Cullis, 1986; Petersen & Gould, 1999).

For the purposes of the WHO/UNICEF Joint Monitoring Programme, rainwater is classified as an "improved" water source. In the same way that groundwater is considered to be safe because of the water-purifying effects of the soil layer, rainwater is perceived to be pure because it has not come into contact with contaminated surfaces. Yet, rainwater carries with it a wide range of chemical pollutants that it dissolves out of the atmosphere. Often the level of these pollutants is insignificant, but in regions that are affected by particularly high concentrations of airborne pollutants the chemical quality of rainwater may be compromised.

Contamination with microorganisms can occur during collection, for example as a result of washing off bird and animal droppings that may be present on the catchment surfaces. Thus, practical measures for protecting the quality of the water include: management of the catchment area; water collection procedures that discard the first flush of water from the catchment surface; and design, cleaning and maintenance of the storage reservoir. Guidance on the design, construction and maintenance of rainwater catchment systems is given by Pacey & Cullis (1986) and Petersen & Gould (1999).

Groundwater sources

Groundwater constitutes 97% of global freshwater and is an important source of drinking-water in many regions of the world (Howard et al., 2006). The main advantage of groundwater over other sources is that it is often of good microbiological quality and may be consumed without treatment. Furthermore, the quality of the groundwater can be protected by applying simple design rules for constructing barriers around the abstraction point. Nevertheless, using groundwater has disadvantages. In particular, abstraction of the water can impose technical and economic constraints on supply projects. In addition, the groundwater in some regions is high in mineral content, for example, nitrate, fluoride and arsenic, in concentrations that may present a significant risk to the health of consumers. The following are common sources of groundwater:

- *Upland springs* have the advantage of being easily protected from contamination often because of their remote location, and they are often at a sufficient elevation to allow for the use of gravity-fed distribution systems.
- *Artesian springs and wells* discharge water under pressure and as a result they are not easily contaminated. A flowing artesian well behaves like an artesian spring. In a non-flowing artesian well, the water level in the well is above the water table, but requires pumping. The yield and the possibility of overdevelopment are of concern, because excessive withdrawal from artesian systems is likely to significantly reduce the pressure in the aquifer.
- *Deep wells* can be located in unconfined or confined aquifers. The groundwater in unconfined aquifers is exposed to the surface, whereas the groundwater in a confined aquifer is covered by an impermeable overlying stratum. Groundwater in an unconfined aquifer is more susceptible to pollution than water in a confined aquifer.
- *Infiltration galleries* consist of free flowing groundwater that is abstracted by means of perforated pipes laid at right angles to the direction of groundwater flow. Infiltration galleries are beneficial in wetlands near coastal areas where the deeper water is saline. The gallery can pick up the fresh superficial water.
- *Shallow wells* are a widely used source of groundwater in many developing countries. Shallow wells can be fitted with a pumping device to abstract water, or the water can be drawn with a simple bucket and rope. The main problem with shallow wells is that they often suffer from deficiencies in both the quantity and quality of the water.

Surface water sources

Surface water sources include large rivers, ponds, lakes and small upland streams which originate from springs or collect run-off from watersheds. The quantity of run-off is dependent on a number of factors, the most important being the amount and intensity of rainfall, vegetation, and the geological and topographical features of the area under consideration. By its very nature, therefore, surface water is likely to have a highly variable quality in terms of both its chemical and microbiological content. However, contaminants can also be quickly dispersed and reduced in concentration by dilution. In order to protect the consumer from possible sudden changes in quality, surface water may have to be treated before it is used. The costs and difficulties associated with surface water treatment, particularly the day-to-day problems of operation and maintenance of water treatment plants, need to be carefully considered before deciding to exploit surface waters. The following are some of the types of surface water sources:

- *Upland streams* offer the best potential for surface supplies. Their watersheds are small and hence relatively easy to protect. Upland streams also offer the potential for developing gravity supplies. However, dry weather flows may be insufficient to meet demand so impoundment may be necessary to provide seasonal storage. Irrespective of the isolation of an upland source, a sanitary survey and quality determinations in wet and dry seasons are necessary to provide a basis for watershed control activities and water treatment.
- *Lakes* can be an excellent source of water. If located at high elevation they provide additional gravity flow. However, the potential for impaired quality as a result of activities on the watershed and on the lake itself are high. Lakes subject to pollution may present more problems than rivers because rivers tend to cleanse themselves after the pollution has abated, while lakes may require long periods of time to overcome the effects of polluting discharges.
- *Rivers* provide convenient sources of supply for small communities. However, they are not the most suitable source because they are often of poor quality as a result of land uses and other river uses. Construction of facilities to extract water from large rivers is likely to be costly, with additional pumping costs if rivers are at a low elevation compared to consumers.

Regional supply

Small systems that supply water to communities have been shown to be sensitive to economies of scale (Sauer, 2005). If a water source is adequate, developing it to double or triple its capacity, as can be achieved through joint enterprise by several communities, can reduce the unit cost substantially. Hence, two or more communities developing a source together may each achieve significant savings, and may make it economically viable to develop a higher quality source. For example, a source at a greater distance could be jointly developed, where development would not be feasible for each community separately.

SOURCE PROTECTION AND SUPPLY INTERVENTIONS

In this section we provide a brief overview of the methods that are available for protecting the quality of a water source while simultaneously providing a means of abstracting and delivering the water to the consumer. Although protection of the water source and the delivery of water to the consumer can be achieved separately, it is helpful to consider the two interventions together in order to provide the maximum benefits to the community.

Access to drinking-water is defined by the WHO/UNICEF Joint Monitoring Programme for Water Supply and Sanitation as the source being less than one kilometre away from its place of use and it being possible to reliably obtain at least 20 litres per member of a household per day. Interventions should first aim to bring the water source closer to the consumer, at least until it meets the criterion for access. In developing countries, the benefits to women and young girls that come from time saved in collecting water and increased personal safety are significant. The short case study described in Box 7.1 shows how improved access was achieved by building a rainwater collection system for the household. But the example also demonstrates the importance of combining improved access with an improved source of water. There can be only limited benefit to the community from providing improved access to an unimproved source.

Box 7.1 Example of the benefits accrued from installation of a rainwater harvesting system

In the dry Eastern African village of Nampuno, Hadija Suleiman and her daughter Fatuma used to walk twice a day the 4 km to the nearest reliable well with good drinking-water. Together they carried the 60 litres the family needed daily. The long trips with heavy loads exhausted them. Fatuma could attend the school only for part of the day. Then they got the roof rainwater catchment. Now, they use rainwater for drinking and cooking, and for their vegetable garden. The surplus vegetables are sold at the market. From that extra income Hadija's husband plans to build an extra rainwater tank.

Source: Smet (2005).

Several excellent publications are available that provide guidance on the planning and construction of improved water supplies (Cairncross & Feachem, 1993; Schouten & Moriarty, 2003; Skinner, 2003). We would recommend referring to at least one of these publications, or a similar text, in order to gain an understanding of the design constraints and the requirements for the construction of each system. In addition, there are several publishing houses that specialize in the publication of guidance documents and manuals for intermediate technology interventions; two examples are the SKAT Foundation (http://www.skat-foundation.org/) and Practical Action (http://practicalaction publishing.org/). You will also find practical inspiration from other organizations such as the IRC International Water and Sanitation Centre (www. irc.nl) and the Water and Engineering and Development Centre at Loughborough University (http://wedc.lboro.ac.uk/).

Resource and source protection

Resource and source protection is achieved by implementing effective catchment management programmes. Catchment management aims to decrease the amount of contamination that enters the water resource, thereby reducing the amount of treatment that is required to supply safe and clean water. Catchment management is critical in the execution of WSPs (WHO, 2004). Two steps are needed: hazard identification; and design and implementation of control measures. Hazards in the catchment may arise from both human and natural factors. It is important that the influence of all factors is understood before effective control measures, including treatment, are considered.

Approaches to catchment assessment and catchment management vary considerably depending on the nature of the catchment and the hazards that it contains. Some examples and case-studies can be found on the United States Environmental Protection Agency web site (http://cfpub.epa.gov/safewater/ sourcewater/). A similar approach to the management of water resources is prescribed in the European Union Water Framework Directive published in 2000. The aim of the Water Framework Directive is to rationalize water policy and legislation in the European Union, to set up water management on the basis of river basin districts. One of its key objectives is: "to provide for sufficient supply of good quality surface water and groundwater as needed for sustainable, balanced and equitable water use". In practice, compliance with the Water Framework Directive requires interventions at the level of land-use protection measures (such as nutrient and soil management, or the introduction of buffer strips), particularly as the land-use protection measures may affect the quality of surface water.

The protection of groundwater sources can be carried out using a similar approach to catchment management. Such protection often falls under the title of groundwater protection zones (in the United Kingdom) or wellhead protection (in the United States). For example, the United Kingdom

Environment Agency has divided groundwater source catchments into four zones (http://www.environment-agency.gov.uk/maps/info/groundwater):

Zone 1 (inner protection zone). Any pollution that can travel to the borehole within 50 days from any point in the zone is classified as being inside zone 1.
Zone 2 (outer protection zone). The outer zone covers pollution that takes up to 400 days to travel to the borehole, or 25% of the total catchment area – whichever area is the biggest.
Zone 3 (total catchment). The total catchment is the total area needed to support the removal of water from the borehole, and to support any discharge from the borehole.
Zone of special interest. This is usually where local conditions mean that industrial sites and other polluters could affect the groundwater source even though they are outside the normal catchment area.

Box 7.2 Catchment-sensitive farming

Catchment-sensitive farming requires a partnership between farmers, regulators and others. Farmers and their advisers need to increase their understanding of pollution risks and prevention. At a problem site they may have to improve the storage of dirty water, and manage soil and chemical use to limit losses of pollutants to water. This approach should be promoted through advice, incentive schemes, and regulation where needed.

Source: Environmental Agency (2007)

Supply interventions

In some rural areas, the most significant – and often the only – intervention is to create a means of capturing, containing and abstracting a water source. A common technique for harvesting rainwater is illustrated in Figure 7.1. Figures 7.2 to 7.7 illustrate some examples of methods used to exploit groundwater and surface water.

Unless the groundwater is emerging at the surface from a spring, the exploitation of groundwater requires two interventions: a means of access; and a means of bringing the water to the surface. The most basic and perhaps the most common method for exploiting groundwater in rural areas, and in many urban areas, is the hand-dug well (Figure 7.2). The well is constructed by digging a shaft into the ground to below the water table. Inflow of surface water into the well is prevented by the construction of a cover and plinth to prevent direct access of water and other forms of contamination from the surface, and by the installation of a suitable lining around the shaft of the well to prevent inflow from sub-surface flow.

Figure 7.1 An example of rainwater collection. *Source*: Carlevaro & Gonzalez (2011)

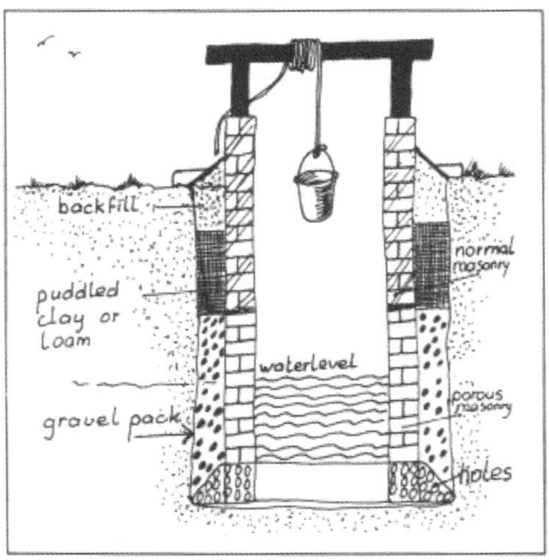

Figure 7.2 A hand-dug well. *Source*: Carlevaro & Gonzalez (2011)

A variety of methods can be used to raise water to the surface. The simplest and cheapest method is to use a bucket and rope. This method is, however, likely to introduce contaminants into the well unless extreme care is taken to protect the rope and the bucket. For example, ropes and buckets can often be seen discarded at the side of the well, in direct contact with soil and other sources of environmental contamination (Figure 7.3). By installing a low-cost and simple pulley system above the well (Figure 7.2), the rope and bucket are protected from immediate sources of contamination, and the microbiological quality of the water in the well will improve (A. Cronin, personal communication, 2007).

Figure 7.3 An example of inappropriate storage of a rope used to draw water from a hand-dug well. *Source*: S. Pedley

The installation of a mechanical pump above the well, such as a hand pump (Figure 7.4) or treadle pump, greatly increases the level of protection of the groundwater by creating a sealed cap over the mouth of the well (a type of wellhead protection). However, the cost of supplying and installing a pump is much higher than providing a bucket and rope, and there are ongoing maintenance costs that need to be considered to keep the pump operating.

Protecting the quality of groundwater emerging from springs requires a different approach to the design of engineering interventions. The challenge is not to provide access to the water, but to provide protection to the catchment and the area immediately surrounding the eye of the spring. An example of an idealized spring protection scheme with a small collection box and distribution system is illustrated in Figure 7.5. The diagram shows how the eye of the spring has been developed by the construction of an enclosed spring box that prevents direct access of contamination to the water emerging from the spring. The spring box is then covered with soil and a drainage ditch is dug above the spring to divert the run-off from upland surfaces away from the soils overlying the spring.

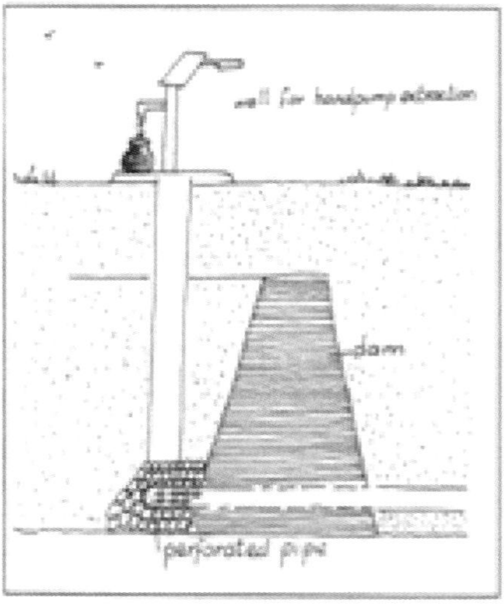

Figure 7.4 Hand pump used to abstract groundwater. *Source*: Carlevaro & Gonzalez (2011)

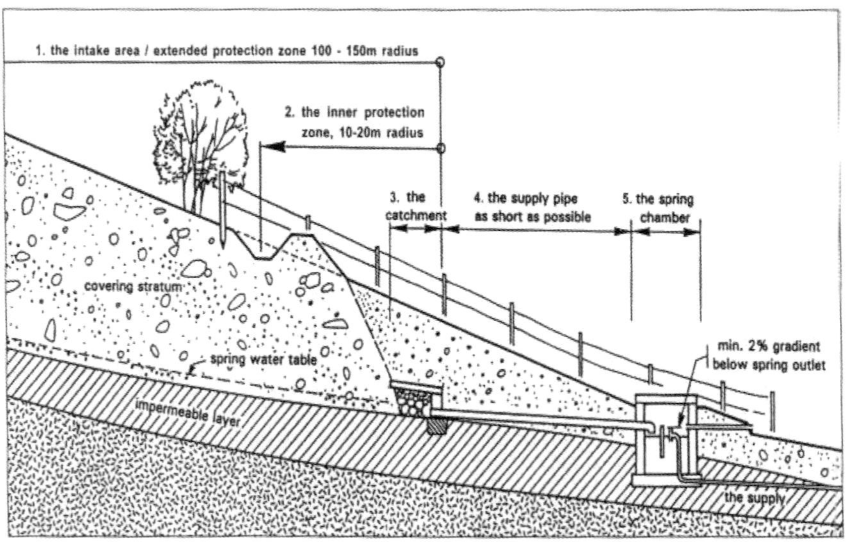

Figure 7.5 Protected spring. *Source*: Carlevaro & Gonzalez (2011)

In summary, provided that the groundwater source is selected with care, and that appropriate construction methods and materials are used to protect the source, groundwater can be of a high quality and may often be used without any treatment. At most, a good groundwater source may require only disinfection to eliminate the potential for pathogens to be transmitted to the consumer. Less protected and more vulnerable groundwater sources may require continuous treatment by disinfection in order to make the water safe.

By contrast, surface water is susceptible to rapid and significant fluctuations in the quality of several parameters, including microbiological parameters, as a result of human and animal activities in the catchment, and environmental factors such as rainfall. Surface water sources such as rivers and streams may also be vulnerable to variation in flow rates over short and long timescales (daily or seasonal variations). To reduce the susceptibility of the water supply to changes in river flow rates and to provide a reserve of water for use during periods of dry weather, reservoirs are often created by retaining the surface flow behind a dam (Figure 7.6). Worldwide there are approximately 800 000 dams, of which approximately 40 000 are considered large dams and over 300 are classified as major dams. The majority of dams, therefore, are relatively small constructions.

Figure 7.6 River impoundment creating a small reservoir. *Source*: Carlevaro & Gonzalez (2011)

While the main objective of constructing a dam is to create a reservoir of water, a potential side-effect of the reduction in the flow of a river is to improve the quality of its water by creating conditions for suspended material to settle out and to carry with it contaminants from the water column. The longer the water remains in the reservoir, the greater the improvement in water quality. However, the reduction in flow rate from impounding the water can introduce hazards that were not present before the construction of the dam. Large bodies of static water will attract wild fowl and encourage recreational use of the water, both of which can be a source of pathogens. Furthermore, studies have reported increased rates of schistosomiasis in regions where dams have been constructed (Hunter et al., 1993).

Techniques for abstracting water from surface water sources can be of a number of types. In the review by Carlevaro & Gonzalez (2011), four different water intakes are described: the protected side intake; the river bottom intake; the floating intake; and the sump intake. One example, the protected side intake, has been included here for the purpose of illustration (Figure 7.7). In this example, the intake provides a stable place in the bank of a river or lake where water can flow into a channel or enter the suction pipe of a pump (Carlevaro & Gonzalez, 2011).

Figure 7.7 Protected side intake for abstraction of surface water. *Source*: Carlevaro & Gonzalez (2011)

The design of the surface water intake has no effect on the quality of the water other than to limit the sediment load. Treatment of surface waters is thus more complicated than the treatment of groundwater because of the need to deal with the wide range of water quality conditions that can be experienced. Therefore, the application of the multiple barrier principle of water treatment – several stages of treatment – is essential if surface water is to be used as a source of drinking-water.

Water treatment

On several occasions in the preceding paragraphs we have referred to the use of water treatment to produce a supply of safe drinking-water from a potentially contaminated source, such as a lake or river. Later in this chapter we discuss water treatment at the household level and describe some of the technologies that are available to households to protect and improve the quality of their water supplies at the point of use. At this point in the chapter it is appropriate that we provide a brief overview of water treatment for piped water supplies.

Apart from the high quality groundwater sources that are free from chemical and microbiological contamination, water that is being delivered through a piped distribution system will require some form of treatment to improve its quality and then to protect its quality in the distribution system. With regard to the latter, even the high quality groundwater sources may receive treatment by disinfection to prevent recontamination of the water by microorganisms that grow on the inner surfaces of pipes in biofilms. Thus, planning the right type of treatment requires a detailed knowledge of the water source.

From abstraction to delivery into the distribution system, the process of water treatment is constructed from a series of stages, each stage producing an incremental improvement in the quality of the water. In general, there are seven stages of water treatment, which are abstraction, chemical dosing, flocculation, settling, filtration, disinfection and pumping into the distribution system.

Chemical dosing, flocculation, and settling have been used in some regions of the world for several centuries, to improve the aesthetic quality of the water by removing visible suspended solids. An added benefit of removing the suspended solids, which was not understood until the early 20th century, is that the processes also remove some of the microbiological contaminants from the water. In the 1700s, filtration was introduced into water treatment as an additional or alternative method of removing particles from the water. Once again, the added benefit of filtration is that it will remove microbiological contaminants from the water. The famous observation of the distribution of cholera cases in Hamburg and Altona during the German epidemic of 1892 remains the most persuasive demonstration of the efficacy of filtration. The following passage is taken from the New York Times (12 March 1893):

> *But Hamburg had nearly 18,000 cases of cholera last year, and 7,611 deaths were reported, while Altona escaped with only 562 cases and 328 deaths. Although the water used in Altona is more thoroughly polluted when it is taken from the Elbe than the water in the same stream at the Hamburg intake, it is carefully filtered through sand before it enters the city pipes. On the other hand, the water used in Hamburg is subjected to no process of filtration. It is to this filtering of the Altona water that the Board of Health of Germany ascribes the comparatively small mortality from cholera in that city. Moreover, it is believed that many of the cases in Altona were imported from Hamburg or were due to the consumption of Hamburg water by residents of the smaller city.*

The use of disinfectants, such as chlorine, as a final stage in the treatment of water was introduced later, following the discovery that microorganisms are capable of causing disease and that these microorganisms could be killed by disinfectants.

Although the stages of water treatment were introduced at different times and in response to different requirements, the resulting process has created a series of barriers to the transmission of pathogenic microorganisms such that the failure in one process does not necessarily compromise the quality of the final product. This is known as the multiple barrier principle and it is an important consideration in the design of water treatment systems.

SMALL-SCALE WATER DISTRIBUTION SYSTEMS

The value of a point water source sited at a distance from the community can be improved considerably by adding a distribution system that can deliver water to standpipes close to individual homes or groups of houses, or to taps inside each house. As well as providing residents with easier access to water, the presence of a tap inside or very close to the house has been shown to be effective in reducing morbidity from diarrhoeal disease (Cairncross & Valdmanis, 2006). Despite the apparent benefits of a water distribution system over a distant point source, the drawbacks of installing a distribution system are significant: cost; the skills required to design and construct the system; equipment; possible legal issues, such as access rights to land; appropriate training and capacity building; and continuing operation and maintenance. Furthermore, as systems get larger there may be a need to review and revise the policies that are in place to introduce support from local authorities and other relevant bodies.

Cairncross & Feachem (1993) recommend constructing rural water supplies that are of a fail-safe character. This may lead to a decision that several tube-wells with hand pumps will be more appropriate than a piped water supply from a distant source. The cautionary note sounded by Cairncross & Feachem (1993) applies both to developed and developing country settings. It emphasizes the need for adequate consideration at the planning stage of a distribution system to the way in which the system will be operated, maintained and financed (Kerr, 1989). A brief introduction to water distribution systems is provided by Skinner (2003).

Water distribution systems vary considerably in size and complexity. Perhaps the simplest would involve collecting water from a protected spring and distributing it to a community by gravity through a single pipe that terminates in a standpipe. At the other end of the scale are the large, zoned distribution networks that supply major population centres. Nevertheless, they all consist of three elements that are interdependent and indispensable: a source of water; the physical works which bring the water from the source to the consumer, treating and storing the water as necessary; and the organization that manages and operates the system (Kerr, 1989).

In the right setting, a system can be designed consisting of a source, treatment, storage and a distribution network that operates entirely by gravity flow. These settings are, however, rare and some type of device will generally be required for pumping water to higher levels or for forcing water through the network of pipes. Whatever type of pumping device is used, it will add a further level of complexity to the system, a higher level of operation and maintenance, and it will render the system vulnerable to closure if it should fail. The capacity and operational characteristics of the pumping device need to be calculated by a competent person who is familiar with the design and performance criteria of the distribution system, including any treatment and storage facilities.

The application of the fail-safe principal recommended by Cairncross & Feachem (1993) continues into the layout of the distribution system. The simplest layout is a branch system, or dendritic system, as shown in Figure 7.8. The distribution main carries the water from the reservoir to the areas of consumption; the service pipes distribute the water to the points of use. The disadvantage of this type of system is that it can lead to stagnant water in the dead-ends and a loss of supply to some areas if there is a failure in the network.

Some of the disadvantages of the branched system can be overcome by connecting together the ends of the service pipes. This simple intervention, sometimes called the gridiron system, allows for a better circulation of water through the system and reduces the potential for water to stagnate in remote or infrequently used parts of the network. An example of how this can be achieved from the basic system is illustrated in Figure 7.9.

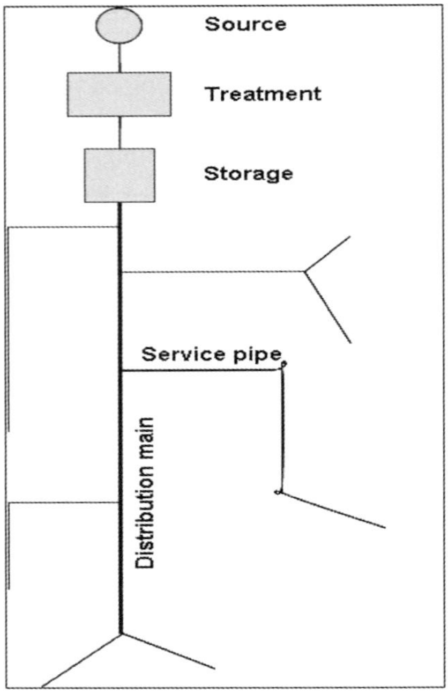

Figure 7.8 A branched (dendritic,) water supply distribution system

Although the gridiron system is an improvement on the branched distribution system, the design in this example is constrained by the original layout of pipes and the resulting network of pipes may not be the most efficient for circulating water, or the easiest to manage in the event of a breakage in the system.

A characteristic of basic gridiron systems, is that the network is supplied by a central feeder pipe. A further enhancement of the gridiron system is to create a loop from the distribution main within the interconnected network of service pipes. This configuration is known as the ring system and has considerable advantages over all other systems. In particular, it allows for good circulation of the water, it is safe in the case of breakdowns, and the supply is not interrupted during repairs.

Water haulage

The distribution of water to consumers by motorized tankers, donkey carts or by hand-pulled carts is common in many countries. Several studies of water haulage in developing countries have shown that between 50% and 80% of the

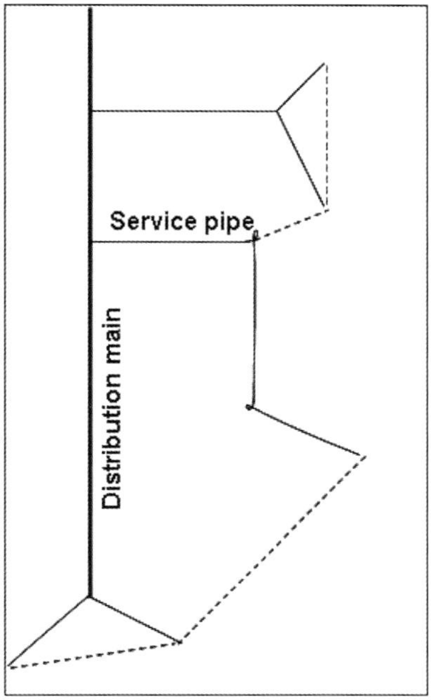

Figure 7.9 Gridiron distribution system

domestic water supply market in urban areas is held by small, independent water providers who transport water from a source to the house (Solo, 1999; Collignon & Vezina, 2000). Water haulers, therefore, are a significant part of the water distribution system even though such services have not been recognized or appreciated by national and international agencies. As a consequence, water haulers in developing countries are viewed with suspicion and distrust, rather than as a potentially valuable resource that could be incorporated into the formal water supply system by the introduction of suitable policy frameworks and supporting regulation. In developed countries, water haulage is also a common method of water delivery to remote households and small communities. In contrast to developing countries, however, the water haulers in many developed countries are regulated, and guidelines are available for the management of water in tankers (for example see Nova Scotia Department of Environment and Labour, 2005).

Household interventions

By and large, people will use household water treatment for two purposes: to improve the safety of the water by reducing the level of harmful contaminants;

and to improve the aesthetic quality of the water by reducing parameters that affect the taste, odour or colour of the water. An example of the latter application of household water treatment is the use of jug and in-line water filters that pass treated water through a granular active carbon filter in order to reduce the taste from residual chlorine, and to partially reduce the hardness of the water. More significant, however, are the household water-treatment systems that are designed to improve the safety of the water. The following paragraphs concentrate on interventions that can be used in households that do not have a piped water supply.

Household, or point-of-use water treatment, and improved water storage practices reduce the bacteriological contamination of water held in the home (Zwane & Kremer, 2007). Furthermore, these interventions, and others discussed below, have been shown to be effective in reducing the incidence of diarrhoeal disease (Clasen et al., 2006; Zwane & Kremer, 2007), although there is evidence to suggest that this response may be influenced by the level of sanitation within the community (VanDerslice & Briscoe, 1995; Esrey 1996; Gundry et al., 2004; Eisenberg et al., 2007). Studies reviewed by Sobsey (2002) show a range of reductions in household diarrhoeal diseases of 6% to 90%, depending on the technology, the exposed population and local conditions. A comprehensive review of household water treatment and water storage technologies prepared by Sobsey (2002) concludes:

> "*The most promising and accessible of the technologies for household water treatment are filtration with ceramic filters, chlorination with storage in an improved vessel, solar disinfection in clear bottles by the combined action of UV radiation and heat, thermal disinfection (pasteurization) in opaque vessels with sunlight from solar cookers or reflectors and combination systems employing chemical coagulation-flocculation, sedimentation, filtration and chlorination*".

All of these systems have been shown to dramatically improve the microbiological quality of water. At least two of them – solar disinfection in clear plastic bottles (heat plus UV radiation) and chlorination plus storage in an improved vessel – have been shown in epidemiological studies to significantly reduce diarrhoeal and other infectious diseases, including cholera. More recently, Clasen et al. (2006) concluded that filtration is the most effective household intervention to improve the microbiological quality of drinking-water and is as effective at preventing diarrhoea as other environmental approaches, such as improved sanitation, hand washing with soap, and improved water supply (Fewtrell et al., 2005).

EDUCATION AND COMMUNITY TRAINING

A significant measure of the success of water and sanitation interventions at all levels is the degree to which the project outcomes are sustained and enhanced after the project has been completed. The experience accumulated over many years by people working in the field has been that the top–down approach of project management, traditionally used for the implementation of water and sanitation projects in developing countries, is seldom successful or sustainable (Sobsey, 2002). The evidence of these programmes – abandoned remains of water and sanitation technologies – can be found in many of the most disadvantaged communities. These relics represent a failure to engage the communities in the planning and implementation of the project, and to incorporate the aspirations of the community into the outputs of the project.

Sobsey (2002) argues that behavioural, motivational, educational and participatory activities are essential elements of any successful and sustainable introduction of water treatment technology. To address this issue, several toolkits have been developed to ensure that community participation is central to the decision-making process. The most widely used and successful of these toolkits is participatory hygiene and sanitation transformation (PHAST), published by WHO (Sawyer et al., 1998). As its basic principle, PHAST makes participation a central theme of learning. This is represented by the self-esteem, associative strengths, resourcefulness, action-planning and responsibility (SARAR) method (World Bank, 1996). PHAST describes a process for community participation using seven steps, with the aim of "helping people to feel more confident about themselves and their ability to take action and make improvements in their communities" (Sawyer et al., 1998). The seven steps are shown in Figure 7.10.

For isolated or remote communities, it may be difficult to develop a dedicated educational programme addressing issues related to developing, managing, using and maintaining drinking-water systems. In these cases, the only feasible solution may be to incorporate a drinking-water component into educational programmes of other sectors. In most countries, the ministries of agriculture run agricultural extension systems. While their initial focus was on providing a link between the agricultural research community and farming communities, over the years such programmes have evolved towards providing education aimed at improving the quality of life of rural communities. In the majority of cases such programmes will contain messages about the management of irrigation water, and any drinking-water messages could be linked to these, particularly in areas where, in reality, the boundaries between different uses of water are fuzzy. Education about establishing, operating and maintaining small drinking-water systems can

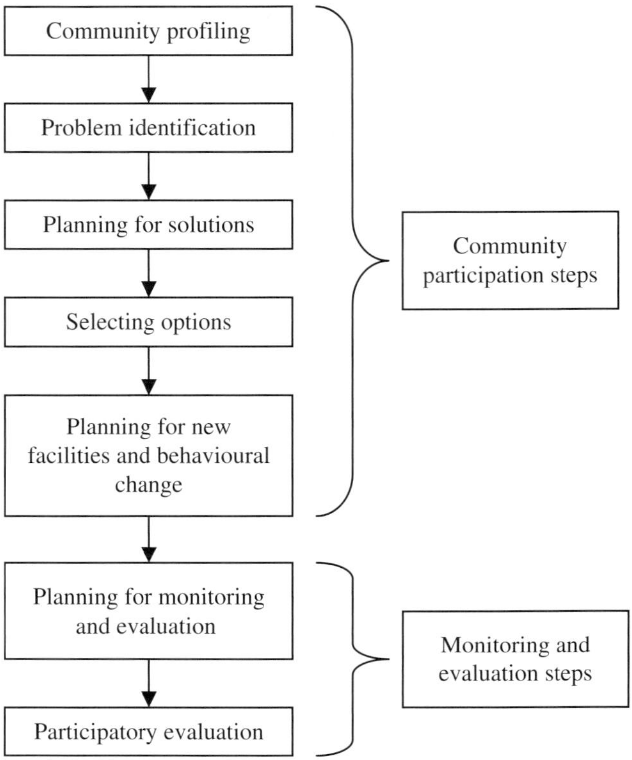

Figure 7.10 The seven steps in the PHAST process

be combined with education about other domestic uses of water, including water supply for livestock and small-scale peri-domestic horticultural activities.

Traditional extension programmes have evolved into more participatory systems over the past ten years, and these so-called farmer field schools deliver programmes that engage farmers in active learning processes about how to sustainably management the natural resources that make up their agricultural production system. Clearly, modules on drinking-water supply and sanitation would fit into farmer field school programmes, as an extension of the already on-going efforts to promote the safe use of wastewater, excreta and greywater in agriculture and aquaculture.

POLICY INTERVENTIONS

The success of technical, social or regulatory interventions to improve water supply is often dependent on the policy environment within which they operate.

It is important, therefore, that appropriate policies are in place at a national and a local level to influence programme managers to select the best option for each community. For example, Howard (2002a) points out that without policies that support the development of improved water supply and which place the emphasis on providing improved services for disadvantaged people, advances in other areas may produce little benefit.

International agencies, governments and local authorities, or their equivalent, play an important role in creating the policy frameworks that will facilitate improvements in water supply for small communities. Policies agreed at the international level can be used to drive improvements to water resource management at the national level, which, in turn, can lead to improvements in water quantity and quality, and the appropriate allocation of water resources between domestic, industrial and agricultural use. The MDGs are a good example of a global commitment that has implications for development and economic policy at the regional and national level.

While international policy is important in many ways for the protection and management of water resources, because it can exert an influence on governance and conflict, it is the policy framework at the national and local level that has more immediate practical implications for the supply of drinking-water to small communities. At the level of the national government, the policy framework will be complicated; policy decisions that have implications for the water sector will be made by different departments. For instance, policies within the agriculture and industry sectors may influence water quantity by determining water rights, permitting water abstraction and controlling discharge. They may also influence the quality of water resources by the discharge of pollutants to air and water. Therefore, it is important that policy analysts and decision-makers in the water sector have an understanding of the policies in other sectors that can affect water resources, and that they can assess the consequences of implementing these policies and associated regulations.

Inevitably, there will be limitations in the information that is available to fill gaps in the conceptual understanding of issues and to inform policy development. At this point, decision-makers should introduce policy frameworks – for example, scientific, health, social or economic – to support activities that will fill the knowledge gaps. For example, tools may need to be developed to help assess the risk to water resources from other activities, and policies and regulations may need to be introduced that require environmental impact assessments of new developments.

One significant knowledge gap in many countries is the burden of disease related to the consumption of water, and how this burden is distributed among the different social groups and water supply systems (see Chapter 5). This information is essential in order to prioritize policy interventions and to measure the impact of these policies. Creating the necessary intersectoral collaboration is vital if data for estimating the burden of disease are to be collected and used effectively. For example, the health sector may be required to collect and report data about waterborne disease outbreaks. Policy decisions will be needed about the types of disease that will notified, the frequency of reporting, the internal and external communication systems, and so on. At the same time, the water sector may be required to identify and report water quality problems to the health sector, so that epidemiological studies can be targeted to strengthening the estimates of disease burden. Not least, policies need to be introduced to set out the actions to be taken in the event of a contamination event and the procedures for communicating with the users.

Another area where national policy is required is in the development of water quality standards and the standardization of analytical methods. As a starting point, many countries have adopted the WHO guideline values for parameters as the national standard for drinking-water. In the light of current thinking about water quality, this policy may be supplemented by the decision to introduce the WSP approach for water management and to use the output from the application of WSPs to revise the initial water quality standards. Within this framework, decisions need to be taken about the individual mandates and roles of institutions in the water sector and in the health sector in monitoring and surveillance.

Policy development at a local level must take place within the context of the national policy framework. Often, the decisions being taken at the local level are comparatively simple, and the policies supporting these decisions may be drafted by technical staff, such as the programme managers installing water and sanitation technologies. For example, the establishment of a local database requires a simple policy decision. But if the local database does not collect, store, analyse and report data in a way that is compatible with the national strategy for data collection, it may have very little overall value. As well as developing policy, programme managers should be in a strong position to use data from their surveillance programmes to lobby for changes to policy or for the introduction at a higher level of new, more relevant policies. Indeed, evidence-based approaches are increasingly being demanded in all sectors of socioeconomic development (Howard, 2002b), and a well-designed surveillance programme will provide the necessary evidence.

The outputs of surveillance data can be used to influence policy-making with regard to water supply improvement. Also, surveillance of water supply systems can be linked to other surveillance programmes, such as disease surveillance. Howard (2002a) discusses policy-making under three main headings: water quality; communal services to disadvantaged people; and source protection, minimum treatment requirements and distribution management. Under each heading, we highlight below the key points raised by the author. Howard (2002a,b) provides a complete discussion of the topic.

Water quality. The WHO *Guidelines for drinking-water quality* recommend that countries establish their own standards for drinking-water quality using parameter values in the guidelines to inform the parameter values in the standards (WHO, 2004). However, passing legislation to regulate drinking-water quality will be ineffective if the policies are not in place to enable the appropriate implementation of the legislation.

Communal services to disadvantaged people. To ensure that services are provided to disadvantaged people, policies may need to include guiding principles about access to water, preferred sources, costs and payment, community participation in decision-making, and sanitation coverage.

Source protection, minimum treatment requirements and distribution management. Policies may be required to cover the basic components of source protection, minimum treatment requirements using the multiple barrier principle, and distribution management. These are critical components in ensuring that water supplies continue to provide high quality drinking-water. Policies promoting impact assessment (including environmental impact assessment and health impact assessment) are essential to ensure that the impacts of development projects on the drinking-water situation and on the health status of affected communities are taken into account at the early planning stages. Development activities planned in a range of different sectors (for example, the construction of a hydropower dam, the development of an area for plantation agriculture, a new mining operation, or the opening up of an area with new roads) may all affect the catchment area for drinking-water systems, the distribution of drinking-water to human settlements, or the management of wastewater flowing out of these settlements. At the same time, drinking-water projects themselves should also be subject to impact assessment procedures. As has been shown in parts of South-East Asia, the introduction of traditional storage jars in an attempt to improve the access of rural communities to safe drinking-water can have unforeseen consequences in areas where dengue fever is endemic. Such storage jars provide a major breeding site for mosquito vectors.

Policies are created and implemented at national and local levels. At both levels, it is important that the policies being implemented adequately address any barriers to higher service levels that have been revealed by surveillance programmes. These barriers typically relate to inadequate distribution infrastructure, poor water-supply management, high cost of connection, high recurrent costs, and poor perception of improved water supplies. In particular, service levels to low-income areas should be addressed.

REFERENCES

Cairncross S, Feachem R (1993). *Environmental health engineering in the tropics: an introductory text*. Chichester, New York, Wiley and Son.

Cairncross S, Valdmanis V (2006). Water supply, sanitation and hygiene promotion. In: Jamison DT et al. eds. *Disease control priorities in developing countries*. New York, Oxford University Press, 771–792.

Carlevaro F, Gonzalez C (2011). *Costing improved water supply systems for low-income communities: a practical manual.* CDROM. Geneva, World Health Organization.

Clasen T F et al. (2006). Preventing diarrhoea with household ceramic water filters: assessment of a pilot project in Bolivia. *International Journal of Environmental Health Research*, 16: 231–239.

Collignon B, Vezina M (2000). Independent water and sanitation providers in African Cities: full report of a ten country study. Washington, DC, World Bank Water and Sanitation Programme.

Deepesh M et al. (2004). Planning and design of cost–effective water harvesting structures for efficient utilization of scarce water resources in semi-arid regions of Rajasthan, India. *Water Resources Management*, 18: 219–235.

Environment Agency (2007). *The unseen threat to water quality: diffuse pollution in England and Wales report.* Bristol, United Kingdom Environment Agency, 18.

Eisenberg JNS et al. (2001). A public health perspective for establishing water-related guidelines and standards. In: Fewtrell L, Bartram J, eds. *Water quality: guidelines, standards and health.*. London, International Water Association Publishing, 229–256.

Eisenberg J et al. (2007). Integrated disease control strategies: balancing water, sanitation and hygiene interventions to reduce diarrheal disease burden. *American Journal of Public Health*, 97: 846–852.

EnHEALTH (2004). *Guidance on use of rainwater tanks.* Melbourne, Australian Government Department of Health and Aging.

Esrey SA (1996). Water, waste, and well-being: a multicountry study. *American Journal of Epidemiology*, 143: 608–623.

Fewtrell L et al. (2005). Water, sanitation, and hygiene interventions to reduce diarrhoea in less developed countries: a systematic review and meta-analysis. *Lancet Infectious Diseases*, 5: 42–52.

Godfrey S, Howard G (2005). Water safety plans. Book 1. Planning urban piped water supplies in developing countries. Loughborough, Water Engineering Development Centre, Loughborough University.

Gundry S et al. (2004). A systematic review of the health outcomes related to household water quality in developing countries. *Journal of Water and Health*, 2: 1–13.

Hanson K et al. (2003). Cost-effectiveness of social marketing of insecticide-treated nets for malaria control in the United Republic of Tanzania. *Bulletin of the World Health Organization*, 81: 269–276.

Howard G (2002a). *Water supply surveillance: a reference manual.* Loughborough, Water Engineering Development Centre, Loughborough University.

Howard G (2002b). *Water quality surveillance: a practical guide.* Loughborough, Water Engineering Development Centre, Loughborough University.

Howard G et al. (2006). Groundwater and public health. In: Schmoll O et al. eds. *Protecting groundwater for health: managing the quality of drinking-water sources.* London, International Water Association Publishing, 3–19.

Howard G, Bartram J. (2003). *Domestic water quantity, service level and health.* World Health Organization WHO/SDE/WSH/03.02. (http://www.who.int/water_sanitation_health/diseases/WSH03.02.pdf, accessed 13 May 2008).

Howard G et al. (2006). Groundwater and public health. In: Schmoll O et al. eds. *Protecting groundwater for health: managing the quality of drinking-water sources.* London, International Water Association Publishing, 3-19.

Hunter JM et al. (1993). *Parasitic diseases in water resources development: the need for intersectoral negotiation.* Geneva, World Health Organization.

Kerr C (1989). *Community water development.* London, Intermediate Technology Publications.

Nova Scotia Department of Environment and Labour (2005). *Portable water hauler guidelines.* Retrieved from http://www.gov.ns.ca/nse/water/docs/PotableWater HaulerGuidelines.pdf, accessed 13 May 2008.

Pacey A, Cullis A (1986). *Rainwater harvesting: the collection of rainfall and runoff in rural areas.* London, Intermediate Technology Publications.

Payment P, Hunter PR (2001). Endemic and epidemic infectious intestinal disease and its relationship to drinking water. In: Fewtrell L, Bartram J eds. *Water quality: guidelines, standards and health.* London, International Water Association Publishing: 61–88.

Petersen EN, Gould J (1999). *Rainwater catchment systems for domestic supply: design, construction and implementation.* London, Intermediate Technology Development Group Publishing.

Rachwal AJ,Holt D (2008). *Urban rainwater harvesting and water reuse: review of potential benefits and current UK practices.* Marlow, Foundation for Water Research: 33.

Sauer J (2005). Economies of scale and firm size optimum in rural water supply. *Water Resource Research*, 41: 1–13.

Sawyer R et al. (1998). *PHAST: participatory hygiene and sanitation transformation series: step-by-step guide – a participatory approach for the control of diarrhoeal diseases.* Geneva, World Health Organization.

Schmoll O et al. (2006). *Protecting groundwater for health: managing the quality of drinking-water sources.* London, International Water Association Publishing.

Schouten T, Moriarty P (2003). *Community water, community management: from system to service in rural areas.* Rugby, United Kingdom, Practical Action Publishing.

Skinner B (2003). *Small-scale water supply: review of technologies.* Rugby, United Kingdom, Practical Action Publishing.

Smet J (2005). WELL fact sheet: domestic rainwater harvesting. *Waterlines*, 24: 13–20.

Sobsey MD (2002). *Managing water in the home: accelerating health gains from improved water supply.* Geneva, World Health Organization.

Solo MT (1999). Small-scale entrepreneurs in the urban water and sanitation market. *Environment and Urbanization*, 11: 117–131.

VanDerslice, J, Briscoe J (1995). Environmental interventions in developing countries: interactions and their implications. *American Journal of Epidemiology*, 141: 135–144.

WHO (2004). *Guidelines for drinking-water quality: recommendations.* Geneva, World Health Organization.

Wilkinson R (1998). What health tells us about society. *Institute for Development Studies Bulletin*, 29: 77–84.

World Bank (1996). Self–esteem, associative strength, resourcefulness, action–planning and responsibility – SARAR. In: *The World Bank participation sourcebook.* World Bank. (http://www.worldbank.org/wbi/sourcebook/sbhome.htm, accessed 26 November 2007).

Zwane AP, Kremer M (2007). *What works in fighting diarrheal diseases in developing countries? A critical review.* Cambridge, MA, Centre for International Development at Harvard University.

8

Estimating the costs of small-scale water-supply interventions

Paul Jagals and Luuk Rietveld

This chapter explains a basic approach to estimating the financial costs of installing, maintaining and operating a small-scale drinking-water supply. The outcome of such estimates for various interventions can then be used, along with estimates of the total benefits expected from water improvements, to select the best intervention for a given target group by comparing rates of return. The choice of the best intervention for a specific community will be made in the context of the livelihood patterns of that particular community, and after assigning shadow prices to certain costs.

Costing is but one step in the economic assessment of a water supply project, and an economic assessment is one element in the set of information likely to be used by decision-makers to select the type of system to implement. The full set of information likely to be needed by decision-makers would include economic, environmental, health, social and technical assessments, and feasibility studies.

© 2011 World Health Organization (WHO). *Valuing Water, Valuing Livelihoods.* Edited by John Cameron, Paul Hunter, Paul Jagals and Katherine Pond. Published by IWA Publishing, London, UK.

The costing method proposed here is intended to provide a financial input into a cost–effectiveness or social cost–benefit analysis of small-scale water supply improvements. In a broader sense, costing is an essential element of any economic analysis that involves modifying financial costs through the assignment of shadow prices to reflect true economic value.

The objective here is to identify the financial costs of a small-scale water supply intervention. The costing method described in this chapter aims at providing an incremental price in present-day monetary terms (year zero) of water supply technology to provide water to a community, against which the derived benefits could be measured. On the basis of estimates of costs and benefits, informed decisions can be made. For information on full economic costing, see Carlevaro & Gonzalez (2011).

The method of social cost–benefit analysis described in this book is for use at a national level, by non-specialists and specialists alike. The method is, however, sensitive to local livelihood patterns. To simplify the presentation, the costing of items that are needed for activities at national level, or that are indirect outcomes at national or local level, are ignored. These include external environmental costs, which arise out of local environmental damage or protection, and opportunity costs, which value the forgone benefits of diverting raw water from productive activities such as agriculture to non-productive activities, for example basic domestic uses (with significant livelihood implications). They also include depletion premiums, which value the loss of water supplies from sources that are difficult to replenish, and the share of overhead costs that are needed to run national regulatory and laboratory facilities.

The approach followed here is to estimate the *use costs* of a water supply intervention, such as costs for construction, operation, maintenance, direct administration and overheads. This chapter provides an insight into the simple and effective costing method that a specialist would use and a non-specialist at national or local level would need to understand. In practice, costing preferably should be done at the level of the service provider, for example a regional water management body or a district authority. Depending on local capabilities, the costing could be done by the end-user target group or a local nongovernmental organization.

Costing a system locally implies that a water service provider (usually a district authority) or a local user group has decided to invest in improved infrastructure (storage and treatment facilities, and a distribution network). After constructing and activating the system, the service provider will then continue to spend money on the system for operation, maintenance, future rehabilitation and administration. Expenditures will also cover training, promotion and education (for example, on the use of energy and chemicals). The service provider will have to make sure that these expenses are covered by some form of income (standing charges, consumption rates or subsidy). The main objective of the service provider might not be to make a profit, but to provide an economically efficient water service that would also benefit the target group by improving livelihoods. The purpose of a cost estimate in this situation is therefore to assist the decision-makers responsible for the provision of services by giving them a reliable estimate of the financial value of providing an improved drinking-water supply.

A useful characteristic of small systems is that they allow for incremental improvement as the target group's needs change over time. Such changes might result from population growth or from a move from standpipe to in-house provision. This may lead to a decision to initially choose the most affordable system (the least-cost system), with the intention to incrementally adapt the system's capacity to fit the needs or financial capabilities of a growing population. A practical costing approach must allow for this incremental costing as well.

The following discussion of the methods and procedures for costing a small-scale water system has been structured into three sections. First, we address the challenges of costing a small-scale water intervention, whether in the form of putting in a system where there was none before, or upgrading an existing system. Second, we outline the various elements of financial costs that will be encountered in the process of costing a small-scale water system. Third, we present a simplified costing approach to reliably estimate a pattern of water system costs. This will allow for costing across time should a system be incrementally adapted.

CHALLENGES OF COSTING A SMALL-SCALE WATER SYSTEM INTERVENTION

Local costing may be done for a new water-supply system or for upgrading an existing system. Costing a new system for an area where there is currently no water-supply system is likely to be complicated, because the costing will take place in the early planning stages, before the actual construction begins to take shape. In particular, the costing of the initial stage of a project to install a water system will potentially be subject to large inaccuracies. One reason for this is that relatively little is known during the initial stage about the configuration of the treatment scheme, construction requirements, and specific local conditions. Another reason is that many things will change during the design process.

As a basis for costing the intervention, three critical and interrelated aspects need to be understood: the type of intervention (the system likely to be installed, see Chapter 7); the layout of the system (the relative positions and elevations of source, storage and pipe network); and the size of the elements in the system. Only then can the costing of the envisaged system commence. While costing can be done in detail to cover these uncertainties, it will require a high level of collaboration between service providers and engineering planners. It is therefore preferable for costing to be done by such specialists. For detailed information, including comprehensive checklists, see Carlevaro & Gonzalez (2011).

Costing the upgrading of an existing system can more feasibly be attempted by non-specialists because there will be less uncertainty.

Estimating the approximate cost of a water supply system usually starts with estimating the investment or capital cost. The service provider should, however, look at the complete picture. Costing should therefore be extended to include recurrent (operation and maintenance) costs. This second component is absolutely vital to predicting what the sustainable operation of the system is going to cost the service provider once the system is built. If the recurrent costs are neglected, the intervention will be short-lived and the benefits often negated before they were accrued. This is discussed in more detail later.

Cost estimation necessarily requires a large number of inputs. In order to simplify the data collection and preparation steps, a three-tier data structure is proposed. The first data category captures the engineering parameters. These would typically include technical specifications, such as pump and motor efficiencies, as well as pipe friction coefficients, which are not likely to vary significantly anywhere in the world. Data for these parameters should be measured by people with an engineering background and with the capacity to provide a sound technical judgement based on experience. The second data category captures the monetary parameters. These would typically include the cost of pumps, pipes, holding tanks, fuel and electricity, as well as the ratio between the costs of labour and materials in constructing the system. These parameters will be fairly constant for any particular economic zone. Once these parameters have been calibrated for a particular region, they can be left unchanged while different water supply systems within the region are analysed. The third data category captures the system parameters. These will typically include the types, diameters and lengths of pipes, the volume of storage tanks, and the number of standpipes. These parameters are unique to each water supply system, and have to be determined on site or from engineering drawings.

ELEMENTS OF COSTING

Costs, by definition, consist of all resources required to put in place and maintain the intervention. These include capital costs (investment in planning, preparing the project site, construction, and purchase of hardware) and recurrent costs (operation, maintenance, and monitoring) (De Moel, Verberk & van Dijk, 2006). The cost of a small water-supply system usually includes capital as well as recurrent costs in each of the usual components of water supply: source, treatment, pumping, storage and distribution. The costing method must be robust and it will need to provide reliable estimates by aggregating sets of physical parts of a water supply activity into a single unit of cost. An example is the estimation of the initial cost of water treatment for a new or improved system. To get started, cost-functions can be used that are based on previously completed projects. Cost-functions typically reflect the design capacity as a variable, for example as cost per cubic metre of treated water. This will give the planner a simple and robust estimate of the cost of water treatment for a village of X number of people consuming Y litres of water per person per day.

Capital costs

The term capital goods is formally defined as meaning the stock of goods which are man-made and used in production (as opposed to consumption). Fixed capital goods (durable goods such as buildings and machinery) are usually distinguished from circulating capital goods (stocks of raw materials and semi-finished goods which are rapidly used up). In accounting conventions, capital goods are usually taken as those with a life of more than one year, such as land, buildings and equipment.

In the context of developing and installing a small-scale water system, the capital costs represent the total costs that are not expected to recur for significant periods of time. These are costs for the preparation and construction of the system up to the moment that the system becomes operational (De Moel, Verberk & van Dijk, 2006). From that moment on, the system must be operated and maintained so that it maximizes the anticipated benefits. The costs of equipment needed for operation and maintenance are also considered to be capital costs. Capital costs can also occur during the operational lifetime of the system. Examples include expansion of the system and replacement of major (high-cost) parts.

Capital costs usually include the costs related to the construction and equipment of the new system. These costs flow from the preliminary studies conducted during the pre-investment (planning) stage. The preliminary studies are concerned with the technical, economic, social, environmental and health aspects of the construction project.

A drinking-water system consists of a variety of fixed (constructed) installations, such as filter units, clear water reservoirs, and pipes. Depending on the size of the system, construction might include office and sanitary facilities for the staff of the new treatment facility, a workshop and maybe a small laboratory with facilities for the maintenance personnel. Besides these costs, the furnishing of staff facilities, workshop and laboratory constitutes part of the capital costs. The project requires equipment, which will be a capital cost, for example items such as pumps and power systems. Materials are needed to complete the construction, including materials bought or acquired by the community or the municipality in the local markets of the country, as well as imported materials.

The workforce for the construction may include specialists, such as engineers, constructors, technical staff, and social scientists. It will also include unqualified workers who will carry out the excavation work, cleaning, and so on. Lastly, artisans will generally be required, depending on the type of work needed.

Other capital costs will be related to the management of the project, and will include administration, coordination, logistics, transport, communications, office costs, private sub-contractors and quality control, as well as any other unassigned costs of the project. Contingency costs are a fixed amount or a percentage of total capital costs included in a project budget to allow for adverse conditions that will add to the basic costs.

A cost which will often be encountered, and which should be seen as part of capital costs, will be the cost of acquisition of land that might be required for components of the system, for example the site of the treatment facility, or land that will be covered by water when a surface source, such as a stream or river, is impounded.

Provision must also be made for overheads and supervision. Once all the capital investment costs have been estimated, their sum will reflect the net construction cost. A contractor might add a surcharge to allow for site establishment, site clearing, supervision, profit, and so on. Such costs can all be allowed for by adding a percentage to the net construction cost. For example, a typical surcharge for contracts in rural South Africa is 25%.

The costs are then added up to determine the total contract cost. For a new water-supply system, the client also has to bear the costs of planning, surveying, soil investigation, possibly exploratory drilling, contract management, quality control, and so on. These design and supervision costs, paid to consulting engineers or borne by the client's own design staff, amount to an additional surcharge (about 25% over and above the total contract value), which must be added to the other costs to finally determine the total project cost.

Recurrent costs

Recurrent costs comprise all expenditures (staff, parts and materials) that are required to keep a system operational and in good condition (maintenance) after its installation has been completed. Depending on the accounting policy of the provider, certain fixed costs, may need to be covered recurrently on an annual basis (De Moel, Verberk & van Dijk, 2006). An example of this would be the creation of a replacement fund, through annual depreciation levies. Monitoring of the system can be seen as an operational function or as a regulatory function to ensure the quality of the water supply to the community. Monitoring has a cost that can be seen as a separate item in the recurrent costs, or as part of operation and maintenance costs, depending on the needs and extent of the system.

An important point in the context of costing the management of small-scale systems is that maintenance costs are often budgeted for annually at the service provider level, which is usually a tier above the local community level. Operational costs are usually budgeted for at the local level.

The maintenance costs cover all costs for the repair and replacement of parts of installations (for example, pumps or wells) within the predicted lifetime of the water-supply system, in so far as these are not included in the operational costs. Effective maintenance is the key to sustainability of a system but it is often neglected, rendering many small-scale systems ineffective not long after their inception.

In general, operational costs are considered to be mostly costs for acquiring and administering consumables, such as energy, process water and chemicals, as well as disposing of waste. Consumables do not include general maintenance materials (such as paint, lubricating oil and tools) because these should be included under maintenance costs.

Fixed costs are costs arising from obligations to finance and operate the system. They include interest, depreciation and replacement, rents, insurance and taxes. Depreciation is a particularly important aspect of fixed costs, because it allows for the build-up of funds to replace a large piece of equipment or parts of the system such as pipes. Depreciation is the way to earn back, from annual income, costs incurred during construction of the system. Depreciation periods for a water system are relatively long. On one hand, the technical facilities (buildings and pipes) should last a long time. On the other hand, there should be income from water sales and subsidies during the entire depreciation period. While the depreciation period should preferably equal the expected lifetime of the water supply system as a whole, depreciation periods are not necessarily the same for all the components of the system. Buildings, machines, distribution network and inventories all have different lifetimes. Therefore, the costs of funding the replacement reserves have to be determined separately for each component. For capital costs that recur within the project period, best practice is to include capital costs in the year when they are incurred. To determine the capital costs an economic depreciation period is assumed. After this period the component might still have a residual value. In that case the net usage of capital must be considered (capital cost minus residual value). In general, however, while the installation might still be technically adequate, redundancy of the installation or high maintenance costs might minimize this residual value.

Lastly, policy development as well as activities relating to national, regional and local monitoring, surveillance and training incur often substantial costs and should also be regarded as and provided for as fixed recurrent costs. These activities are required to continuously assess and maintain the quality of the service, including protection of the source water, as well as during and after treatment and distribution. They require skilled human resources, laboratory facilities and training facilities, vehicles and sampling equipment. Some of these activities might require an initial capital investment, such as on-site monitoring systems or the cost of a meeting of stakeholders to consider and prioritise new and extended systems. A significant cost component of all these activities can be travel costs. Travel might be required to and from monitoring points, remote training sessions and facilities. These costs might exceptionally include the costs of monitoring and assessing the process of livelihood changes attributable to the intervention – social behavioural change – along with the more common costs of education and promotion. Another cost component that might occur is the cost of water corruption. This reflects an activity where people illegally gain access to the distribution of the water supply (for example by means of illegal connections without a meter). Inequity in distribution is another form of water corruption.

ESTIMATING COSTS FOR A SMALL-SCALE WATER SYSTEM

Three reasons for costing exist for small-scale water supply interventions. A service provider might want to: (a) conduct direct costing; (b) estimate the costs as part of a cost–effectiveness analysis; or (c) estimate the costs as part of a social cost–benefit analysis. We now discuss the third option – estimating the costs as part of a social cost–benefit analysis. For detailed guidance on direct costing, see Carlevaro & Gonzalez (2011). The approach described by Clasen et al. (2007) can be followed to estimate intervention costs for a cost–effectiveness analysis.

Estimates for cost–benefit analyses need not be as detailed as the estimates for a cost analysis or a cost–effectiveness analysis. They can be simple *unit costs*, as shown in Table 8.1. The unit cost approach provides flexibility when a service provider wishes to estimate whether investments to install a new system are more cost-beneficial than investments to upgrade an existing system.

Table 8.1 Typical unit costs for rural water supply systems

	Capital investment (US$ per person)	Recurrent (% annual cost)	System lifetime (years)	Water demand (litres per person per day)
House connection	92–144	20–40	30–50	80–120
Standpost	31–64	0–10	10–30	50–80
Handpump on drilled well	17–55	0–10	10–30	20–30
Dug well	21–48	0–10	10–30	20–30
Rainwater	34–49	5–15	10–30	20–30

Source: WHO/UNICEF (2000)

Unit costs are robust cost estimations of a system. They include *capital costs* as well as *recurrent* costs. Data can be obtained from local sources, in particular from country-specific cost summaries of previously installed schemes.

In the next section, we first discuss the costs incurred by *preliminary requirements* to developing and installing a system. We then consider each of the *activity costs* usually included in a small-scale system. Lastly, we provide a summary (Table 8.1) of unit costs and briefly discuss the calculation of unit costs.

Preliminary requirements

This section aims to describe the full costing process that eventually allows for a unit cost (cost per volume unit) to be estimated.

Costing can begin only after the details of the physical system are known. For planned systems, an inventory has to be developed to the point where the specific components of the system have been clearly identified, for example pipe lengths and diameter, position and sizes of storage tanks, and so on. For existing systems, the convenient option would be to find the original technical drawings and specifications to which the system had been built. This option, however, is often not available. The drawings might be deposited in some remote archive, and it may be difficult, if not impossible, to retrieve them. Even if the drawings are retrieved, care must be taken to compare the details of the original plans with those of the current system, to establish whether the plans have not already been changed since the original construction work.

It is highly recommended that costing should be preceded by thorough fieldwork, in close collaboration with the local community. In the absence of engineering drawings, the most feasible way is to locate and map the system components (which is becoming increasingly easy with Global Positioning System technology), to locate the pipe routes and water connections, and also to assess the quality of the system in terms of maintenance and reliability.

Activity cost estimation

Developing the source

Small-scale water systems are often supplied from groundwater or from perennial protected springs. Because of its inherent characteristics, groundwater in rural areas is often considered safe enough to be provided directly, without treatment, using a handpump. Costs are lower than for other forms of supply, which makes this a popular choice with service providers.

Where there is no other option than to use surface water, construction of impoundments in rivers and streams is usually required to provide, throughout the year, a continuous supply of raw water for treatment and distribution (see Chapter 4). The costs of creating an impoundment to serve a small-scale water-supply system can be a considerable proportion of the whole cost of the system.

The capital costs of groundwater sourcing are twofold: the direct cost of gaining access to an aquifer, either by drilling a borehole or digging a well; and the cost of lining the borehole or well, where it has to penetrate soft material in the earth. A good estimate of drilled-well costs can be made, for example, by using unit rates for linear metres of hole drilled and lined, respectively. The unit cost here is usually the capital cost per metre drilled, including the final finishing of the well, such as the casing and concrete surface collar – depending on the extent of the service rendered by the drilling company. The final capital cost will therefore depend on the depth of the drilled well.

The maintenance cost will be a percentage of the civil structure, as discussed below. The operation costs for the well itself will be minimal if the well was properly installed. Operational costs related to pumping are discussed below.

Costs of surface-water sourcing will mostly be incurred by the creation of an impoundment, as well as by securing the land that the impounded water might cover, the land required for the sourcing activity such as a pumping station, and often the treatment facility. Capital costs can be estimated as the cost per cubic metre of concrete in the dam wall, per running metre of the dam wall or per cubic metre of water stored. The latter would usually be used if the activity required the purchasing of land. A maintenance cost will be required for ensuring the integrity of the impoundment wall, as well as for whatever sluices or valves or other mechanical water outlets there might be. The maintenance cost will be a percentage of the civil structure costs. The operation costs will be incurred by running the system and will often comprise only personnel costs.

Storage

After sourcing, water usually needs to be stored, either for direct distribution or pre- and post-treatment distribution. These activities require a storage tank, which is usually a capital cost item. Three common storage tank types are in use for small-scale water supply systems. The smallest systems generally use prefabricated glass-fibre tanks if and when these are available. Glass-fibre tanks are sized in multiples of about 2500 or 5000 litres up to a maximum size of about 20 000 litres. For storage of volumes larger than about 20 cubic metres, tanks of reinforced concrete might be used. Tanks assembled from prefabricated panels of galvanized steel are also popular because of their ease of construction, and they are available in sizes similar to those of plastic tanks.

Treatment

When water is obtained from a surface-water (and sometimes a groundwater) source, treatment is required. Depending on the quality of the source, simple chlorination may be sufficient. When water is polluted with suspended solids and pathogenic microorganisms, more advanced treatment is necessary, including coagulation and flocculation, as well as filtration. Most treatment items are capital costs incurred in installing the treatment system. These costs depend on the degree of pollution of the source, the number and type of treatment steps, and the scale of the treatment. The larger the scale of the treatment is, the lower the costs per cubic metre of building area. Unit costs for different treatment steps can be obtained from projects that have previously been implemented in similar settings. Part of the capital cost at the treatment site is the installation of a small laboratory for the analysis of water quality. Other capital costs are related to building a secure place for the storage of chemicals, as well as pumping stations and reservoirs. Although the capital costs of treatment are normally not high compared to the capital costs of transport and distribution, treatment requires considerable operation and maintenance. The operation and maintenance costs consist mainly of salaries for operators and laboratory personnel, and the costs of chemicals (such as aluminium sulphate and chlorine) to be dosed during treatment. Water will be lost during the cleaning and backwashing of filters, and the disposal of the resultant sludge must be organized. The loss of water (which can amount to up to 5–10% of the water produced) represents an economic value; and the sludge must be treated before disposal, which also represents an economic and environmental value. These costs must therefore be included in the operation and maintenance costs.

Distribution

Water can be distributed through a pipeline; mobile units such as tanker trucks or animal-drawn carts; and, containers that people in communities use to move water from the supply point and store at home. Costing a distribution system is discussed here in the context of costing these three systems.

Pipelines are usually capital cost items. The cost components of a pipeline consist of the costs of pipes, couplings and shut-off valves. There are also the earthworks needed to excavate pipe trenches, bedding for laying the pipes on, backfilling the pipeline trench after laying the pipe, and labour. For the smaller diameters of pipes used in small systems, the capital costs are about constant and mostly independent of the pipe diameter. Maintenance costs are normally incurred to maintain valves. Operational costs will be incurred to fix major breaks and minor leaks in pipelines.

Mobile distribution might also require considerable capital investment, depending on the type of system. For example, it may require investment in the truck or cart and the animals. The maintenance costs will be incurred in keeping the vehicles and tanks in good mechanical order. Animals have to be kept healthy, which will incur a cost. Vehicle fuel and animal feed are operational costs.

A container-based distribution system requires the purchasing of the containers (a capital cost), and keeping the containers free from dirt and biofilm (a recurrent cost item). These costs can be considerable for a low–income household and should be considered when attempting a cost–benefit analysis. The idea is that an intervention must be optimally effective at a minimum cost.

Costs that are often overlooked when assessing a small system will be those related to the inevitable water losses, especially through distribution. The characteristics of system losses need to be established, and can be seen as operational or other costs.

Pumping

Pumping is an integral part of many small systems across the globe. Whether water is pumped from the source to the treatment works or to the distribution system, pumps have certain characteristics that will enable the costing of the pumping component to be correctly attributed. These characteristics are best determined with the help of a technician or engineer with specific knowledge in this field.

Pump suppliers can provide an estimation of the capital as well as the recurrent costs if they can be provided with information on the net power delivered by the pump. The net power is derived from the static head, an estimate of the friction head, as well as the pumping rate if it is known. The pumping rate can be estimated from the pipe diameter and assuming a pipe flow velocity (typically between 0.6 and 1.0 metres per second for small diameter pipelines). From this, the size of the motor to drive the pump can be derived. Such a motor can be electric, but in rural areas would usually be a fuel-powered motor, which has implications for the recurrent costs.

Public source points

In a small-scale system, the community will often source its drinking-water supply from the taps at the end of standpipes. The standpipes are connected to the distribution pipeline. The capital investment involves the taps, pipework and connecting fittings, which have a nominal size of 15 mm, 20 mm (the most common) or 25 mm, the latter being the most sturdy. To facilitate the filling and lifting of containers, most taps are installed as part of a small concrete platform, with the vertical pipe encased in some form of concrete pedestal. The maintenance of the taps has proven to be a substantial recurrent cost, in that the tap is often not designed for heavy use.

General remarks on estimating maintenance cost

If the project has a long lifetime, more parts of the installation are likely to be replaced within this period, resulting in higher maintenance costs. The planning (and concurrent costing) of maintenance should identify all the activities involved, and the activity levels for implementation such as hours of work by activity, replacement parts and repairs procedures. The activities and activity levels that typically would be encountered during maintenance may evolve over time, making it possible to estimate an annual cost for maintenance as an annual constant cost equivalent to the present value of the changing maintenance costs over the lifetime of the equipment. A more straightforward, generic method is to estimate the maintenance costs per year as a percentage of the construction costs. The civil, mechanical and electrical parts require maintenance to different extents, requiring different percentages. These specific percentages have to be estimated as accurately as possible, together with the appropriate depreciation periods.

One important element that may make it difficult for a service provider to use the straightforward approaches described above is the growth of the population served by a new system – especially in developing countries. The system would require constant upgrading, even though a major extension might not be undertaken. To keep an eye on effective and sustainable maintenance, the service provider would use a monitoring measurement such as water demand.

The total demand to be met by a water system is a critical parameter which, in a way, drives the entire cost estimate, but is especially important in planning and costing maintenance. In an existing system total demand can be directly measured from a bulk flow meter, and this value should take precedence. When this is not possible (a bulk flow meter is seldom present or working), it may be possible to determine the pumping rate (by volumetric measurement of how rapidly the storage tank is filled, or simply by reading the information plate on the pump) and determining for how many hours a day the supply pump would typically work. Failing this, the water production has to be estimated from the consumer end by multiplying the per capita water demand by the population size. The per capita demand can be estimated by counting the containers filled at a typical standpipe, and the population size either from census data (where available) or by counting the households and estimating the average occupancy per household on some demographic basis.

To the estimated water demand must be added the water lost through leaks in the pipes and at the connections. This is measurable by checking the night flow, but this measurement is less than reliable when standpipes are left open during the night for irrigation or other purposes. A preliminary estimate can be made by assuming the values for leakage given by Farley & Trow (2003). However small the water loss may seem at first sight, it is important to allow for some leakage. For spread out rural systems with low demand, leakage may be significant.

General remarks on estimating operational costs

In general, operational costs are constants over time if the price of inputs and activity levels or the volume of drinking-water delivered remain constant. In this case, operational costs can be estimated as a constant annuity over the life of the equipment. If this is achievable, an annual constant equivalent cost could be estimated, in the same way as for maintenance costs. Operational costs normally include consumables such as electricity, treatment chemicals and liquid fuel for pumping stations. The projected consumption of these items, as well as their prices, are readily estimated.

The estimation of personnel costs is much more difficult, as staff are often only employed part-time in a small-scale water supply system. Small systems may only require one hour of operation per day (to stop and start a pump). Often a specific person will be given the task along with other community duties such as waste collection. In other instances, one person will be responsible for the operation of more than one small system, to which must be added the extra cost and time of moving between these systems. It is clear that no single algorithm could capture all these permutations. There is no option but to estimate the personnel costs from first principles.

The same arguments hold for the cost of equipment required for monitoring and maintenance. A certain minimum of laboratory equipment, for example, is required for monitoring, whether hundreds of samples or only a few samples have to be analysed per week. Often, the monitoring will be performed by a better equipped regional laboratory to obtain some economy of scale, but at the expense of having to transport samples.

Estimating unit costs

Unit costs will vary between countries and will depend on the initial investment (capital) costs, the recurrent costs, the lifetime of the system and the water demand (the water requirements per person per calculation period). Table 8.1 contains typical figures for these cost components, derived from WHO/UNICEF (2000) and Haller, Hutton & Bartram (2007). Here the capital investments are given in US$ per person. Other indicators could also be used for determining capital investment. For instance, a house connection is usually associated with a fixed cost per household irrespective of how many people are in that household; for a well, the cost is often estimated at a fixed rate in US$ per metre of well dug.

We take house connections as an example to demonstrate the method for deriving unit costs. We assume a capital cost of US$ 120, linear depreciation at 2.5%, and a system lifetime of 40 years. Assuming an interest rate of 7.5%, the fixed costs for a house connection in a small rural water system will be $(2.5 + 7.5) = 10\%$ of US$ 120, which is US$ 12 per person per year. The recurrent costs will be approximately 30% of US$ 12, which is US$ 3.6 per person per year, amounting to an annual cost of US$ 15.6 per person. Assuming a demand of 100 litres per person per day, the total annual demand is 36 500 litres which is 36.5 m^3. The unit cost will then be US$ 0.43 per m^3.

POLICY, EDUCATION AND COMMUNITY TRAINING FOR COST ESTIMATION AND RECOVERY

In the previous paragraphs a comprehensive explanation of the different costs of all water supply components is given. Ultimately, these costs have to be recovered in the form of physical payment. Generally in developed countries the consumer pays directly for these costs. The water price per cubic metre is often equal to the total costs per cubic metre (including fixed costs, maintenance and operation costs, as well as profits to be paid out). In developing countries the water price often appears to be much lower than the real cost of water supply, and sometimes the water is even supplied free of charge. This does not mean, however, that it is meaningless to calculate the costs for construction, maintenance and operation. On one hand, cost estimation is important for comparison of different options for interventions. On the other hand, national, provincial and local government must determine, in coordination with the service provider, the need for subsidies to keep the system running and to invest in future systems.

The practice in many developing countries is not to cover the needs for operation, maintenance and new interventions from the recovered costs (by payment and subsidies). As a result the condition of the system deteriorates, leading finally to increased water losses, interruptions in water supply and increase of consumables. Leakage can also lead to deterioration of the water quality. In addition, in time the water demand will grow (as a result of population growth). With an increase in interruptions in the supply, consumer satisfaction will go down, leading to less willingness to pay. Once this situation is reached, it will be hard for the service provider to move matters without substantial investment. Good costs estimates and realistic water pricing and subsidies are thus of the utmost importance.

To be able to make good cost estimates, it is recommended that not only the service provider, but also national, provincial and local government keep record of former investments. From this database, the characteristic costs for elements of a water supply system can be derived. It will then be possible to determine costs for new interventions more accurately, taking account of specific local conditions.

Good cost estimates will also enable service providers to set realistic budgets for operation, maintenance and new investments on an annual basis. In regard to the procurement of spare parts for the water supply system (taps, valves, pipes, spare pumps), it is important to consider not only the price of the parts, but also their lifetime, the need for maintenance and the service provided by the suppliers. For example, imported taps may be cheap to, but it might be preferable to obtain good taps on the local market.

Cost estimation and cost recovery at the level of the service provider, as well as at local, provincial and national level are essential. Therefore, special courses and workshops should be organized to create awareness among policy- and decision–makers, and consumers. Such awareness creation will also help service providers to streamline activities and provide good maintenance and operation.

Cost estimation and cost recovery are fundamental elements in keeping a water supply system running and allowing maximum access to potable water for consumers. However, they are clearly not the only elements. There should also be sufficient technical capacity to construct, maintain and operate the systems.

In addition, logistics are of major importance. For example, if it takes one week to repair a pump, the community will not be supplied by the system during that week. Consumers may be forced to use unimproved sources, leading to increased health risks. This would negate all efforts to ensure accurate cost estimation and recovery.

REFERENCES

Carlevaro F, Gonzalez C (2011). *Costing improved water supply systems for low-income communities: a practical manual.* CDROM. Geneva, World Health Organization.

Clasen T et al. (2007). *Cost–effectiveness of water quality interventions for preventing diarrhoeal disease in developing countries.* London, IWA Publishing: 599–608.

De Moel PJ, Verberk JQJC, Van Dijk JC (2006). *Drinking water: principles and practices.* London, World Scientific Publishing.

Farley M, Trow S (2003). *Loss in water distribution networks.* London, IWA Publishing.

Haller L, Hutton G, Bartram J (2007). Estimating the costs and health benefits of water and sanitation improvements at global level. *Journal of Water and Health*, **5**: 467–480.

WHO/UNICEF (2000). *Global water supply and sanitation assessment 2000 report.* Geneva, World Health Organization and United Nations Children's Fund.

9

Estimating health impacts of interventions with a focus on small-scale drinking-water interventions

Helen Risebro and Paul R. Hunter

Having established a framework for identifying and costing interventions, we now move to the benefits side of economic assessment. In this chapter the focus will be on estimating the health impacts of small-scale interventions giving improved access to safe drinking-water for a target group of people. The method could also be applied to other environmental health interventions such as new sanitation facilities or the introduction of educational or behavioural change programmes. Interventions can be conducted at the individual, household or community levels. They can include, for example, educating children about the importance of handwashing, digging pit latrines in people's homes or building

© 2011 World Health Organization (WHO). *Valuing Water, Valuing Livelihoods*. Edited by John Cameron, Paul Hunter, Paul Jagals and Katherine Pond. Published by IWA Publishing, London, UK.

small-scale water treatment plants collectively. Most of the discussion will be around how to estimate the likely change in disease prevalence and incidence that would follow a particular intervention. However, any environmental intervention is likely to have a broader impact than its effect on disease occurrence. This broader impact will be dealt with in Chapter 11. The present chapter will consider both the effect on disease burden and the livelihood impacts that are primarily mediated by the expected reduction in disease.

In order to estimate the impact of an intervention on disease burden, we need to know how common the disease in question is currently (see Chapter 5) and the likely impact of the proposed intervention on disease occurrence (what proportion of illness will be prevented).

The expected disease reduction is then given by:

$$\text{Expected reduction} = \text{Current frequency} \times \text{Expected impact}$$

The expected reduction in disease burden can then be derived from the expected reduction in disease frequency multiplied by the average disease burden per case.

As suggested in Chapter 5, probably the best estimates will come from studies conducted in the area of the proposed intervention. However, few analysts will be fortunate enough to have access to such primary data. Rather, they will have to rely on previous studies or global estimates. The options available to the analyst are: to conduct an intervention study in the area under investigation; to rely on previous studies of interventions in similar situations (or, preferably, on the conclusions of systematic reviews of such studies); to use risk assessment; or to rely on global estimates of disease. We briefly describe these approaches below, giving primary data where available. Finally, we consider phased interventions as well as the special situations where chronic disease is a major effect of lack of access to safe water or where a substantial proportion of the population is particularly vulnerable.

INTERVENTION STUDIES

When attempting to estimate the impact of an intervention by means of a primary study, the analyst should consider several questions:

- Is it necessary to conduct a new study or is there sufficient information already available to enable the analyst to estimate the likely health impact?
- What resources are available in order to conduct a study, bearing in mind that certain study designs can be very expensive?
- How urgently are the results likely to be needed?

- What is the nature of the proposed intervention? Is it likely to have an impact at the individual, household or community level?

The impact of interventions can be studied using a variety of study designs, including cross-sectional, cohort and randomized controlled trials. Intervention studies can be conducted under natural conditions (accidental trials, such as outbreaks), under uncontrolled conditions (public measures, such as the introduction of a new water treatment plant), or under controlled conditions (clinical trials or field studies) (Payment & Hunter, 2002).

For small-scale interventions, such as the installation of a new water treatment plant for a village, a cohort or cross-sectional survey may be most appropriate and would certainly be less costly than a randomized controlled trial. As discussed in Chapter 5, in the cohort approach the investigator follows up a group of people over time to determine whether they develop illness or how frequently they become ill. In the cross-sectional survey, people within the target population will be interviewed at a single point in time to determine the prevalence of disease. In either the cohort or cross-sectional study design, it is important to survey the target group that has received the intervention and one or more comparable control groups yet to receive an intervention. A control group is a group that has had no intervention but is studied to compare with the group having the intervention. Differences in illness between the study and control group are assumed to be due to the intervention. The incidence or prevalence of illness can then be compared between the groups and, assuming other things to be equal, any difference can be attributed to the intervention.

The double difference evaluation approach combines both longitudinal (a study design where information is collected from the same people over time) and cross-sectional (a study design where information is collected from people just once) data, but is very resource demanding. Even with comparable control groups, the two big problems are attribution and the assumption that all other things are equal. It is often the case that the wealthiest village with most political muscle gets the new water treatment plant first; the people in this village would be expected to be healthier anyway. Some of these biases can be overcome if the communities under study are surveyed for a period before the intervention as well as after.

The randomized controlled trial is one of the most robust epidemiological study designs and permits simultaneous comparison of outcomes in a group of individuals. Study participants are randomly assigned to one or more intervention groups where the intervention is expected to influence disease status, or to the control group which receives either the status quo or a placebo (sham) intervention. Randomized controlled trials overcome many of the problems with bias that affect other study designs, though they are not without their problems. Not least of these is the cost involved in conducting randomized controlled trials, and the challenge of "blinding" people to whether they have the real or sham intervention.

When an intervention requires participant involvement, effectiveness can also be assessed through observation or self-reports of participant compliance, knowledge and acceptability. For example, in a cluster randomized controlled trial of household-based water treatment in rural western Kenya, field workers used questionnaires in the 5th and 15th week of the study to assess knowledge of and attitudes towards the intervention (Crump et al., 2005). As another example, a cross-sectional study assessed sustainability of changed hygiene behaviour in rural areas of India using a short questionnaire, spot observations of household environment and latrine, "pocket voting" (confidential voting to declare normal hand-washing practice) and a demonstration of hand-washing technique (Cairncross et al., 2005).

Observable indicators of behaviour change, for example unanticipated spot checks on filter use, are more objective than self-reported measures, and thus preferable. In a comparison of questionnaires with direct observations to measure hygiene practices in rural Zaire, mothers were found to over-report "desirable" behaviours yet open questions led to underreporting of certain behaviours (Manun'Ebo et al., 1997).

USING PREVIOUS STUDIES AND SYSTEMATIC REVIEWS

It may be the case that a relevant study with the intervention of choice has been done in the area of interest, or at least in a similar area. This would allow the use of such a study to estimate the potential health impact of any proposed intervention. In most instances, however, no single study will be available. In this case, it would be best to use results from a systematic review and meta-analysis. There have been several recent systematic reviews and meta-analyses of water, sanitation and hygiene interventions in developing countries. Some of the results of these studies are given below. As yet, very few similar studies have been reported from high-income countries.

Table 9.1 includes the results of several meta-analyses undertaken to assess the effect of water, sanitation and hygiene interventions on health outcomes (e.g. diarrhoeal disease incidence). Meta-analysis combines results from similar studies to provide a pooled estimate. In the example in Table 9.1, a relative risk of less than 1 indicates reduced diarrhoea associated with the intervention.

Table 9.1 Pooled estimates of relative risk (random effects) of diarrhoea for water, sanitation and hygiene interventions

Intervention	Number of studies	Random effects, mixed age cohorts[a] RR (95% CI)	Probability of heterogeneity (X^2)	Consistency (I^2)	Health outcome	Setting	Type of study included	Reference
Hygiene	**11**	**0.63 (0.52–0.77)**	**0.01**		Diarrhoea, dysentery	Less developed countries with endemic illness	Intervention	(Fewtrell et al. 2005)
Excluding poor quality studies	8	0.55 (0.40–0.75)						
Handwashing	5	0.56 (0.33–0.93)						
Education	6	0.72 (0.63–0.83)						
Sanitation (latrine provision)[b]	**2**	**0.68 (0.53–0.87)**	**0.24**		Cholera, diarrhoea			
Water supply (new source, supply or connection)	**6**	**0.75 (0.62–0.91)**	**0.01**		Diarrhoea, typhoid, cholera			
External source only	4	1.03 (0.73–1.46)						
Household connection	2	0.90 (0.43–1.93)						
Standpipe or community connection	3	0.94 (0.65–1.35)						
Water treatment and storage	**15**	**0.69 (0.53–0.89)**	**0.01**		Diarrhoea, dysentery, giardiasis, cholera			
Source treatment only	3	0.89 (0.42–1.90)						
Household treatment only	12	0.65 (0.48–0.88)						
Household treatment (excluding poor quality studies)	8	0.61 (0.46–0.81)						
Household treatment – rural	6	0.61 (0.39–0.94)						
Household treatment – urban or periurban	5	0.86 (0.57–1.28)						
Multiple interventions	**5**	**0.67 (0.59–0.76)**	**0.02**		Diarrhoea, dysentery			

(Continued)

Table 9.1 (*Continued*)

Intervention	Number of studies	Random effects, mixed age cohorts[a] RR (95% CI)	Probability of heterogeneity (X^2)	Consistency (I^2)	Health outcome	Setting	Type of study included	Reference
Water treatment, storage and supply	**8**	**0.50 (0.41–0.61)**[c]	**0.0001**	**85.7**	Diarrhoea	Countries with endemic illness	Randomised & quasi-randomised controlled trials	(Clasen et al., 2007)
Source-based	1	0.45 (0.43–0.47)	NA	NA				
Household-based	7	0.49 (0.36–0.65)[c]	0.002	71.8				
Household-based chlorination	4	0.41 (0.26–0.65)	0.003	78.4				
Household-based filtration	2	0.41 (0.21–0.79)[c]	0.66	0				
Household-based improved storage	1	0.79 (0.61–1.03)	NA	NA				
Household-based chlorination	**10**	**0.71 (0.58–0.87)**[a]	**0.001**	**76**	Diarrhoea	Less developed countries	Intervention	(Arnold & Colford 2007)
Excluding poor quality studies[d]	8	0.66 (0.51–0.87)[a]						
With safe storage and education	7	0.65 (0.46–0.80)[a]						
Without safe storage and education	3	0.87 (0.62–1.22)[a]						
Urban or periurban	3	0.63 (0.50–0.80)[a]						
Rural settings	5	0.89 (0.71–1.13)[a]						
Handwashing	**20**[e]	**0.57 (0.46–0.72)**[f]	S		Diarrhoea, dysentery, shigellosis, cholera, typhoid	Developed and less developed countries	Intervention, case-control, cohort, observational	(Curtis 2003)
Excluding poor quality studies[d]	6	0.58 (0.49–0.69)[f]	NS					

Hygiene (good quality studies)[d]	6	0.67[f]	Diarrhoea	Worldwide	Intervention and observational	(Esrey et al. 1991)
Sanitation	**11**	**0.78**[f]				
Excluding poor quality studies[d]	5	0.64[f]				
Water quantity and quality	**29**	**0.78**[f]				
Excluding poor quality studies[d]	7	0.81[f]				
Water quality	**7**	**0.83**[f]				
Excluding poor quality studies[d]	4	0.85[f]				
Improved water and sanitation conditions	**7**	**0.80**				
Excluding poor quality studies[d]	2	0.70				

[a]Relative risks from Arnold & Colford (2007), based on people aged less than 18 years.
[b]Improved water supply and hygiene education provided in conjunction with sanitation.
[c]Includes studies with multiple intervention arms compared with a single control.
[d]As defined by authors.
[e]Number of data points.
[f]Relative risk obtained from Fewtrell et al. (2005).
NA, not applicable; NS, not significant; S, significant.

Large heterogeneity between studies may reduce the validity of meta-regression models. Where available, the results of statistical tests for consistency (the fraction of the variation in the effect estimate caused by heterogeneity) and probability of heterogeneity are included in Table 9.1. These results are intended to inform the reader of the reliability of the estimates of relative risk; high consistency and significant heterogeneity indicate poor reliability.

Significant heterogeneity can be further explored using meta-regression. Results not reported in Table 9.1 include seven studies of the effect of water treatment or storage on diarrhoea (Gundry et al., 2004). Meta-regression of six of these studies with significant heterogeneity (between-studies variance of 0.095) revealed that the proportion of households in the trial with adequate sanitation was significantly associated with greater effectiveness of the intervention.

If health outcome datasets are not available, it is possible to assess the effect of water and sanitation interventions on water quality. Table 9.2 presents the results of a review by Arnold & Colford (2007). Wright, Gundry and Conroy (2004) conducted a systematic review of microbiological contamination between source and point of use in developing countries (the study is not included in Table 9.2). The quality of drinking-water was found to decline significantly after collection in many settings. Significant variability existed between studies; meta-regression revealed that contamination after collection is proportionately greater where faecal and total coliform concentrations in source water are low. In addition, the percentage of point-of-use samples contaminated with faecal coliforms was lower where households covered water containers.

So, for example, in Table 9.1 the relative risk of diarrhoea in households after use of a household-based filtration system is given as 0.41 (95% CI 0.21–0.79). This implies that in households in low-income communities with unfiltered water, the expected reduction in illness as a result of the intervention would be around 59%, and the actual reduction would be within the range 21–79% (on 95% of occasions). If the actual level of illness in the community was five episodes of illness per person per year before the intervention, the expected reduction would be 5×0.59. So 2.95 episodes of illness per person per year would be prevented (95% CI 1.05–3.95 episodes).

RISK ASSESSMENT

Chapter 5 discusses the role of quantitative risk assessment in determining the disease burden attributable to drinking-water. For many interventions, it may be possible to estimate effectiveness in removing pathogens (Smeets et al., 2006). Most data available apply to large-scale water treatment plants, though it should

Table 9.2 Pooled estimates of relative risk (random effects) of water quality for water, sanitation and hygiene interventions

Intervention	Number of studies	Random effects, <18 years old, RR (95% CI)	Probability of heterogeneity (X^2)	Consistency (I^2)	Water quality outcome	Setting	Type of included study
Point-of-use chlorine drinking-water treatment	**10**	**0.20 (0.13–0.30)**	**0.001**	**91**	*E.coli* contamination of stored water	Less developed countries	Intervention
Randomized controlled field trials	8	0.21 (0.14–0.32)					
Chlorine, safe storage and education	8	0.19 (0.12–0.31)					
Chlorine alone	2	0.25 (0.19–0.32)					
Rural	5	0.21 (0.12–0.38)					
Urban or periurban	5	0.18 (0.10–0.33)					

Source: Arnold & Colford (2007).

be possible to obtain reasonably good information on household interventions. Quantitative microbial risk assessment works best for single pathogens. It should, however, be possible to do quantitative microbial risk assessment on two pathogens, and estimate the overall impact on diarrhoeal disease.

As already mentioned, probably the most commonly available indicators of water quality are indicator organism counts. Although total coliforms do not predict ill-health in humans, there are now many studies that have shown that faecal indicator bacteria (faecal coliforms or thermotolerant coliforms), especially *E. coli* or faecal streptococci, are associated with diarrhoeal disease risk (Hunter, 1997). Although the exact relationship between illness and *E. coli* count varies, such studies may give an indication of the potential reduction in disease risk from some forms of intervention.

For chemical contamination, the relationship may be clearer. For example, if a community suffers a given level of disease because of arsenic contamination of their drinking-water, then this should give the amount of disease preventable by removing arsenic until the supply complies with WHO guidelines. However, estimates become less precise for phased reduction in contamination levels or when the route under investigation is just one of the sources of a toxic chemical (as is the case with lead).

GLOBAL SCENARIOS

WHO has developed various global scenarios to describe the risk of water-related ill–health (Prüss & Havelaar, 2001). These scenarios are necessarily imprecise for an individual country, but they give a reasonably reliable estimate of ill-health. If an intervention moves a community from one risk scenario to another, then that community would experience a different disease burden and the difference in the estimated disease burdens would indicate the value of the intervention on disease reduction.

PHASED INTERVENTIONS

So far, we have considered how to estimate the impact of a single intervention, for example providing safer water through a standpipe or using in-house chlorination. The reality is, however, that for most water supplies in rural settings globally, a single intervention is unlikely to be sufficient to bring the water up to the highest standards – those expected by people living in urban settings in developed countries.

Estimating the impact of multiple and sequential interventions is slightly more complex than estimating the effect of a single intervention. Obviously, sequential interventions are not simply cumulative. For example, household-based chlorination may reduce diarrhoeal illness by 59% (relative risk 0.41) and handwashing may reduce diarrhoeal illness by 44% (relative risk 0.56). Obviously, the two together cannot lead to a reduction of 103%. The most probable reduction in illness would be multiplicative. So, for household-based chlorination alone the reduction would be 59%. The introduction of handwashing would reduce that lower level of illness by 44%. Table 9.3 provides a worked example.

Table 9.3 Worked example showing how to estimate the impact of multiple interventions

	Relative risk	Calculation	Illness rate per person per year	Reduction in illness rate
Before any intervention			5	0
Household-based chlorination only	0.41	0.41 × 5	2.05	2.95
Handwashing only	0.56	0.56 × 5	2.80	2.20
Both household-based chlorination and handwashing	0.41 × 0.56	0.41 × 0.56 × 5	1.18	3.82

The message is clear. Whatever your starting point, the first intervention is likely to have more impact on reducing disease burden than any subsequent intervention. This could explain the relatively disappointing additional impact of multiple interventions or sequential interventions over single interventions, as illustrated in Table 9.3.

Many communities do not have sufficient resources to fund the immediate development of high quality water and sanitation systems that would comply with standards for urban settings in developed countries. Significant health benefits may be obtained from simple one-step interventions. An incremental approach to water safety is implicit in the water safety plan method (WHO, 2004). Rather than setting out to achieve compliance with some standard in one go, the community should identify the intervention that would provide the greatest cost–benefit and then implement it in line with what is affordable. Once that intervention has been implemented, then there can be a further reassessment of priorities and possibly a further intervention.

CHRONIC DISEASE

Estimating the impact of water supply improvements on chronic disease poses different problems from those for acute microbial disease (see Chapter 5). Chronic disease can persist for many months or years. In affected communities, chronic noncommunicable disease may well be the most important contributor to disease burden.

Estimating the impact of water-supply improvements depends on whether water is the predominant or only exposure factor, or one of several. If water is the main factor, then providing an appropriate intervention is sufficient to remove most adverse effects. The health gain will simply be equivalent to the disease burden in the community. This would be the case, for example, for removal of arsenic from drinking-water. If water is just one of several factors, then the calculation would be more complex, and would depend on defining the various exposure routes and their contributions to overall exposure. This is likely to require contributions from experts beyond those normally brought together for water and sanitation studies.

VULNERABLE GROUPS

There are many possible definitions of vulnerability and susceptibility (Balbus et al., 2004). For the purposes of this discussion, a vulnerable group is considered to be composed of people who may be at increased risk of suffering an illness or may have more severe outcomes should they be affected. When considering the impact of waterborne disease on vulnerable groups, it is important to distinguish between those vulnerabilities that are part of the normal life-cycle (youth, pregnancy, old age) and those that are not (HIV/AIDS, cancer or other diseases). Perhaps the most obvious example of the latter is the case of HIV/AIDS and cryptosporidiosis (Hunter & Nichols, 2002).

It is our view that vulnerabilities attributable to the normal life-cycle always have to be taken into consideration in any risk assessment, and that water supply provision should be appropriate for these vulnerable groups. For other vulnerabilities, it may be decided that adequate general environmental standards cannot be achieved within the boundaries of reasonable expenditure and that alternative personal protection measures are needed. Nevertheless, in any intervention aimed at the general population, vulnerable individuals are also likely to benefit. Indeed, it is possible that the value per person of the benefit for a vulnerable individual may greatly exceed that for a non-vulnerable member of the community. If a community is host to a large number of particularly vulnerable individuals (such as communities with particularly high HIV/AIDS prevalence), then consideration should be given to valuing the benefits to this population as well.

CONCLUSION

This chapter provides a framework for calculating changes in the number of episodes of illness in a target population attributable to a drinking-water intervention. If the duration of each episode of each identified illness is known, then the total number of additional days available for activity can be estimated. In addition, the cost of treating each episode can be ascertained using the same survey techniques and the financial saving to the target population can be estimated. This mixture of time and financial savings can then be incorporated into the livelihoods analysis to estimate the pattern of new activities that would be induced. Values can then be attributed to these activities.

In most circumstances it would be adequate to estimate the potential impact based on a prior knowledge or estimate of disease burden and to estimate the actual impact from prior intervention studies or systematic reviews. If interventions are planned within a country, then consideration should be given to measuring the impact of the interventions on health and livelihood. The results of these studies can then be used for future cost–benefit analyses. For regional and global analyses, it is probably satisfactory to estimate impact based on prior global estimates.

REFERENCES

Arnold BF, Colford JM (2007). Treating water with chlorine at point-of-use to improve water quality and reduce child diarrhoea in developing countries: a systematic review and meta-analysis. *American Journal of Tropical Medicine and Hygiene*, **2**: 354–364.

Balbus J et al. (2004). Defining susceptibility for microbial risk assessment: results of a workshop. *Risk Analysis*, **24**: 197–208.

Cairncross S et al. (2005). What causes sustainable changes in hygiene behaviour? A cross-sectional study from Kerala, India. *Social Science and Medicine*, **61**: 2212–2220.

Clasen T et al. (2007). Interventions to improve water quality for preventing diarrhoea. *British Medical Journal*, **334**(7597): 782.

Crump JA et al. (2005). Household based treatment of drinking water with flocculant-disinfectant for preventing diarrhoea in areas with turbid source water in rural western Kenya: cluster randomised controlled trial. *British Medical Journal*, **331**: 478–481.

Curtis V, Cairncross S (2003). Effect of washing hands with soap on diarrhoea risk in the community: a systematic review. *Lancet Infectious Disease*, **3**: 275–281.

Esrey SA et al. (1991) Effects of improved water supply and sanitation on ascariasis, diarrhoea, dracunculiasis, hookworm infection, schistosomiasis, and trachoma. *Bulletin of the World Health Organization*, **69**: 609–621.

Fewtrell L et al. (2005). Water, sanitation, and hygiene interventions to reduce diarrhoea in less developed countries: a systematic review and meta-analysis. *Lancet Infectious Diseases*, **5**: 42–52.

Gundry S et al. (2004). A systematic review of the health outcomes related to household water quality in developing countries. *Journal of Water and Health*, 2: 1–13.

Hunter PR (1997). *Waterborne disease: Epidemiology and Ecology*. Chichester, Wiley.

Hunter PR, Nichols G (2002). The epidemiology and clinical features of cryptosporidium infection in immune-compromised patients. *Clinical Microbiology Reviews*, 15: 145–154.

Manun'Ebo M et al. (1997). Measuring hygiene practices: a comparison of questionnaires with direct observations in rural Zaïre. *Tropical Medicine and International Health*, **2**: 1015–1021.

Payment P, Hunter PR (2002). Intervention studies. In: Hunter PR, Waite M, Ronchi E, eds. *Drinking water and infectious disease: establishing the links*. Boca Raton, CRC Press, p. 191–196.

Prüss A, Havelaar A (2001). The global burden of disease study and applications in water, sanitation and hygiene. In: Fewtrell L, Bartram J, eds. Water quality, guidelines, standards and health – risk management and management for water-related infectious disease. London, International Water Association Publishing, pages 43–59.

Smeets P et al. (2006). *Efficacy of water treatment processes*. MICRORISK Project. http://www.microrisk.com/uploads/microrisk_efficacy_of_water_treatment_processes.pdf (accessed 16 May 2008).

WHO (2004). *Guidelines for drinking-water quality: recommendations*. Geneva, World Health Organization.

Wright J, Gundry S, Conroy R (2004). Household drinking water in developing countries: a systematic review of microbiological contamination between source and point-of-use. *Tropical Medicine and International Health*, **9**: 106–107.

10

Cost-effectiveness analysis in practice

Chris Edwards[1]

We are now in a position to pull costs and benefits together and reflect on how they should be valued and how interventions can be compared.

This chapter looks at how cost–effectiveness analysis is conducted and how it has been used in the health sector. Then we look at its promotion by the World Bank in the 1990s through the Disease Control Priorities Project. The Disease Control Priorities Project is associated with the development of the major physical indicator used by WHO in cost–effectiveness analysis. This is the disability-adjusted life year (DALY) – and we devote a section of this chapter to look at what DALYs are and how they are used. We then look at the system of

[1] The contribution of Tom Clasen in writing an earlier version of this chapter is gratefully acknowledged. However, Chris Edwards retains responsibility for any errors or omissions.

© 2011 World Health Organization (WHO). *Valuing Water, Valuing Livelihoods*. Edited by John Cameron, Paul Hunter, Paul Jagals and Katherine Pond. Published by IWA Publishing, London, UK.

generalized cost–effectiveness analysis promoted by WHO in the late 1990s, and its application to water improvements.

The focus on DALYs as an effectiveness indicator facilitates the smooth transition from cost-effectiveness analysis to social cost-benefit analysis through the valuation of time. But any physical indicator of health improvement could be used as an effectiveness indicator in a cost–effectiveness analysis – including any reduction in episodes of illness or symptoms of ill-health. Reduction in episodes of diarrhoea is a clear candidate indicator in the case of drinking-water interventions and is discussed before we move on to DALYs.

ECONOMIC ASSESSMENT – THE STEPS TO BE TAKEN

Moving from simple costing through cost–effectiveness analysis to social cost–benefit analysis increases the sophistication of economic assessment used to inform investment decisions. We are looking for investments that produce the biggest surplus of benefits over costs for society as a whole, and the maximum gain in overall well-being per unit of expenditure. So far we have covered four steps in the analysis, as follows:

Step one: describe the present position

The purpose of this first stage is to establish the present situation and collect information that will enable the benefits from possible improvements to be estimated. We want to be able to estimate the difference between the situation *with* an improvement and a situation *without* an improvement.

The analysis of the present position is vital in order to assess the costs of alternative improvements and the benefits likely to arise from these improvements. The definition of the current situation – in terms of epidemiology and livelihoods– is discussed in Chapters 5 and 6 and summarized in Table 10.1.

Step two: identify the feasible improvements

The second stage (after identifying and mapping the existing situation) is to identify the feasible interventions to improve access to safe drinking-water (see Chapter 7). The improvements can consist of a single intervention or a package of interventions. Facilities considered improved by the WHO/UNICEF Joint Monitoring Programme are summarized in Table 4.2.

Step three: estimate the costs of improvements to water supply

The third stage is to estimate the costs of each of the feasible improvements. This will involve the estimation of one-off (or capital) costs and recurrent (or operation and maintenance) costs. The structure of financial costs is discussed in Chapter 8 of this book. Hutton, Haller & Bartram (2006) spread the investment costs over the lifetime of the improvement at an interest rate of 3% per year, as shown in Table 10.2. We can treat the capital (investment) costs as they do, or we can list the costs on a year-by-year basis with the capital costs entered as and when they are incurred at or near the start of the project.

Table 10.1 Economic benefits arising from water and sanitation improvements

Beneficiaries	Health benefits from avoiding diarrhoeal disease	Non-health benefits
Health sector	Reduced expenditure on the treatment of disease Value of fewer health workers falling sick	
Non-health sector	Reduced expenditure on the treatment of disease Reduced expenditure on transport in seeking treatment Reduced time lost in seeking treatment Value of time savings resulting from a reduction in illness Value of reduction in deaths	Value of time savings arising from better access to water or sanitation Value of leisure activities relating to better access to water Benefits arising from higher net output in agriculture, industry or services from better water supply (for example, from greater crop irrigation)

Note that the costs in Table 10.2 are in US dollars at year 2000 prices. Because of inflation, it will be necessary to update the costs to prices at the present time, or value all the benefits at year 2000 prices. In any case, it is essential to use the same base year for the prices of all inputs and the values of all outputs. Note also that the costs are estimated per person reached, *not* per household.

All the prices and values may, of course, be in the domestic currency, not in US dollars. The global data values presented in Table 10.2 are estimates using the US dollar as the indicator currency. Numbers should always be rounded and should never have more than three digits. Rarely is any greater precision warranted, given the likely measurement and sampling errors.

Step four: estimate the health benefits

Chapter 9 discusses the estimation of physical health benefits arising from water and sanitation improvements. Table 10.2 summarizes the assumptions made by

Table 10.2 Estimates of the cost of an episode of diarrhoea as an effectiveness indicator

	Variable	Data values	
		Range	Median
Expenditure on the treatment of diarrhoeal disease	Unit cost of treatment per visit (US$)	4.5–21.9	
	Unit cost of treatment per day (US$)	18.3–86.6	
	Proportion of cases hospitalized (%)	8.2	
	Days per hospitalized case	3–7	5
	Outpatient visits per case	0.5–1.5	1
Health workers falling sick with diarrhoea	Information not available from sources quoted below		
Expenditure on the treatment of diarrhoea	Patient costs per day in hospital (US$)	1.0–3.0	2
	Patient costs per outpatient visit (US$)	0.25–1.0	0.5
Transport expenditure in seeking treatment	Percentage of patients using transport	0–100	50
	Costs per visit (US$)	Not available	
Time in seeking treatment:			
of adults	Days off work per episode	1–4	2
	Cost of time (GNP per capita)[a]	0.3–1.4	
of school-age children (5–14 years)	Absent days per episode	1–5	3
	Cost of time (ratio of GNP per capita)[a]	0.3–1.4	
of infants (0–4)	Days sick	3–7	5
	Cost of time	50% of GNP per capita	
Deaths (future years discounted at 3% per year)	Years lost: age 0–4 years	9.5–29.1	16.2
	Years lost: age 5–14 years	15.2–33.8	21.9
	Years lost: age 15+ years	16.3–22.7	19.0
	Years valued at GNP per capita	0.3–1.4	

Sources: Hutton & Haller (2004) and Hutton, Haller & Bartram (2006).

[a] The range of values of time given by Hutton, Haller & Bartram (2006) is from 30% to about 140% of GNP per capita.

Hutton, Haller & Bartram (2006) on the costs of diarrhoea as an effectiveness indicator. The values in Table 10.3 may be used as guidelines, but note that they are in year 2000 prices.

Table 10.3 An example of cost–effectiveness analysis

Years	Actual				Discounted		
	Capital costs (US$ million)	Recurrent costs (US$ million)	Infant deaths avoided (million)	Discount factor at 3% per year	Capital costs (US$ million)	Recurrent costs (US$ million)	Infant deaths avoided (million)
1	140	0	0	0.971	135.94	0	0
2	120	20	10	0.943	113.16	18.86	9.43
3		20	20	0.915		18.3	18.3
4		20	30	0.888		17.76	26.64
5		20	30	0.863		17.26	25.89
6		20	30	0.837		16.74	25.11
7		20	30	0.813		16.26	24.39
8		20	30	0.789		15.78	23.67
9		20	30	0.766		15.32	22.98
10		20	30	0.744		14.88	22.32
Total	260	180	240		249.10	151.16	198.73

HOW COST-EFFECTIVENESS ANALYSIS IS CARRIED OUT

If the outcomes of projects can all be converted into monetary terms, then we can use an approach that compares the benefits to the costs, and prioritizes projects on the basis of social cost–benefit analysis. This approach is looked at in detail in Chapters 11 and 12. Here, however, we look at an approach which is used where the benefits cannot easily be measured in monetary terms, or where valuation data would be expensive to collect or likely to be controversial in interpretation. This approach is cost–effectiveness analysis.

Cost–effectiveness analysis is a method that consists of defining the objectives of a project and choosing the solution that minimizes discounted capital and recurrent costs for a given output or maximizes the output for a given cost. In health terms, the objective will usually involve a specific intervention whose cost has been estimated. An intervention is a deliberate attempt to improve health by reducing the risk, duration or severity of a health problem and the term usually refers to an activity undertaken by a health system rather than by an individual (Jamison et al., 2006a). Activities aimed at promoting behavioural change (for example to reduce the risk of sexually transmitted infections) are, however, also interventions.

It may be that the costs and attributable physical outputs are spread out over a period of time. If this is the case, then they will have to be discounted to the present. Discounting is a standard approach used in economics to give value to time (see Box 10.1).

Box 10.1 Discounting (taking account of time) in cost–effectiveness analysis

Because the costs and benefits of almost any investment project are normally spread over a number of years, a method is needed for putting costs and benefits which occur at different times on a comparable basis in order to obtain a single figure which indicates the profitability of a project.

One dollar receivable in one year's time is not worth as much as a dollar today because if a person had a dollar today, he or she could invest it to obtain a larger sum than a dollar in a year's time. Also, whether or not the money is invested, people prefer to have things now rather than in the future.

Discounting is simply a way of taking account of time by attaching lower weights to cash flows (or benefits) which occur in the more distant future than to those which occur relatively soon. How does it work?

Assume that you have 100 dollars to save in a bank deposit account and that this gives you a rate of interest of 5% per year. In a year's time your \$100 will be worth \$105. This is simply \$100 $(1 + (5/100)) =$ \$105.

Discounting is simply the reverse process. Instead of taking a given sum and seeing how much it will be worth in a year from now, we start with some *future amount* and calculate the equivalent amount today. So if we expected to receive \$105 in a year from now, we could ask what this would be worth today at a discount rate of 5% per year. This is known as the *present value* of the future amount. The discount rate in a year from now at 5% per year is 1 divided by $(1 + (5/100))$ or $1/1.05 = 0.9523809$. Multiply \$105 by this and we get \$100, which is the present value of the \$105 at 5% per year.

Similarly, \$100 invested over two years at 5% per year will give you \$110.25. The discount factor for year 2 at 5% per year is simply 1 divided by 1.1025 or 0.907029. If we multiply \$110.25 by this discount factor (0.907029) we will get \$100. Thus \$100 is the present value at 5% per year of \$110.25 receivable in 2 years from now. There are published tables in which you can find discount factors for different interest (discount) rates and different years, see for example, Gittinger (1973).

In some countries, services such as health and education are provided solely by the public sector and there is no corresponding free market for these services, so there may be considerable difficulties in putting a monetary value on the benefits or outputs. At the same time, we will still want to make the best use of resources. Cost–effectiveness analysis is commonly used where it is felt that a monetary value cannot be placed on the output. The cost–effectiveness ratio will then be the discounted costs divided by the discounted physical outputs, a process which is illustrated in the following example.

Assume that we have an investment in an immunization programme in the health sector, with the initial and recurrent costs as shown in Table 10.2. The objective of the programme is to reduce infant deaths. Therefore the number of infant deaths avoided is shown in Table 10.3. To keep the example more manageable, the table shows the costs and outputs over a period of 10 years only. In reality, the analysis would usually be carried out over a period of 20 years. In the *WHO guide to cost–effectiveness analysis*, Tan-Torres Edejer et al. (2003) argue that cost–effectiveness analysis should cover a period of 10 years after full implementation, which would mean a time-horizon for the analysis of around 15 years.

Using the example in Table 10.3, if we compare the total (undiscounted) costs of US$ 260 million + US$ 180 million = US$ 440 million with the infant deaths avoided (240 million), the cost per infant death avoided is US$ 1.83. The cost per infant death avoided on a discounted basis (using an annual discount rate of 3%) is slightly greater, at US$ 2.01. The cost per death avoided is greater on the discounted basis because the costs are nearer to the present than the deaths avoided. Thus, as emphasized in Box 10.1, the discounting process reflects the value of time.

This cost–effectiveness ratio of US$ 2 per infant death avoided can then be compared with the cost–effectiveness ratios of other health programmes to guide the choice of programme priorities.

There are a number of steps in cost–effectiveness analysis. The first is to define the objective and decide on the indicator of output. In the above example, this was infant deaths avoided. The next step is to estimate the year-by-year capital and recurrent costs of the project. These costs should all be measured on the basis of the price level in a particular year (year 0 in Table 10.3) and outputs also estimated on a year-by-year basis. The final step is to decide on a discount rate and draw up a table like Table 10.3 and do the calculations, finishing with a cost–effectiveness ratio. This cost–effectiveness ratio can then be used as a guide to the relative worth of projects or programmes that claim to reduce infant deaths.

The cost–effectiveness ratio can be calculated on an average basis or on an incremental basis. The incremental cost–effectiveness of interventions will often vary with the level of service coverage (see Jamison et al., 2006a, pages 276 and 277). Thus the cost of reaching the first 1% of the population may be quite high and may yield relatively few health gains. But, as the coverage increases, the average cost may fall and health gains may increase, resulting in a substantial improvement in the cost–effectiveness of reaching an additional group. Then, once coverage is high, reaching the remaining, marginal, segments of the population may again be quite costly. As a result, reaching the marginal population may give a very high cost–effectiveness ratio. This means that the case for reaching the marginal population may have to be decided on the grounds of income distribution, not on the basis of the cost–effectiveness ratio. The case for an average cost–effectiveness ratio is that it can ignore the existing mix of health interventions, many of which may be inefficient and worth scrapping. We come back to this below, in the discussion of generalized cost–effectiveness analysis.

Cost–effectiveness ratios are a useful guide to the worth of different interventions in the health sector. They have been strongly promoted by the Disease Control Priorities Project, which we look at briefly in the next section.

THE DISEASE CONTROL PRIORITIES PROJECT AND DISABILITY-ADJUSTED LIFE YEARS

In the late 1980s, the World Bank initiated work to inform decision-making on setting priorities for the control of specific diseases and to generate cost–effectiveness estimates for health interventions in developing countries. In 1993, this work, which involved WHO and health specialists from across the world and was based on the use of disability-adjusted life years (DALYs) as indicators, was incorporated in the first edition of *Disease control priorities in developing countries*. This was a companion book to the 1993 *World development report*, which focused on investing in the health sector (World Bank, 1993).

In April 2006, the Disease Control Priorities Project released the second edition of *Disease control priorities in developing countries* (Jamison et al. 2006a). This second edition included updated information about the global burden of disease attributable to tobacco, alcohol and psychiatric disorders, which account for an increasing proportion of deaths.

The work of the Disease Control Priorities Project is published in two volumes with a heavy emphasis on cost–effectiveness analysis in both volumes. Useful reading on cost–effectiveness analysis is referred to in Chapter 3 of the first volume and in Chapters 2 and 15 of the second volume.

To ensure consistency, the individual chapter authors adopted, where possible, a common cost–effectiveness indicator: US$ per disability-adjusted life year (DALY) averted. The concept of the DALY had been developed in the late 1980s and early 1990s to provide a broad community health measure that goes beyond just deaths avoided. Thus DALYs go beyond a classification of individuals as either alive or dead. They incorporate standards of health on the basis of disease and disability weights established by WHO. As a result "a DALY measures not only the additional years of life gained by an intervention but also the improved health that people enjoy as a consequence" (Jamison et al., 2006b). In the health sector DALYs have become a widely-used effectiveness indicator.

To ensure consistency in cost estimates in the two volumes, all authors measured costs in US dollars and counted the cost of implementing interventions, but not costs incurred on patients and their families.

A major outcome of the Disease Control Priorities Project was the estimation of cost–effectiveness ratios for health interventions. Table 10.4 summarizes some of these cost–effectiveness ratios expressed in terms of costs (in US$) per DALY averted by interventions aiming to reduce mortality among children under 5 years of age.

Table 10.4 Reducing mortality rates among children under 5 years of age: costs per DALYs averted in the interventions studied by the Disease Control Priorities Project

Intervention	**Cost per DALY (US$)**	**DALYs averted per US$ million spent**
Improved care of children under 28 days old (including resuscitation of newborns)	10–400	2 500–100 000
Expansion of immunization coverage with standard child vaccines	2–20	50 000–500 000
Adding vaccines to the standard child immunization programme	40–250	4 000–24 000
Switching to the use of combination drugs against malaria where there is resistance to current inexpensive drugs (sub–Saharan Africa)	8–20	50 000–125 000

Source: Jamison et al. (2006a).

Table 10.4 reflects the considerable differences in the costs per DALY averted. The Disease Control Priorities Project claims that cost–effectiveness ratios are helpful in making better use of limited health resources, even though there are considerable problems in measuring the costs per DALY averted. As stated on page 272 of *Disease control priorities in developing countries* (Jamison et al. 2006a) "Cost–effectiveness provides the clearest simple way to promote value for money in health; hence the emphasis on it here".

DALYs continue to be widely used both as a measure for estimating the global burden of disease and as an outcome measure for use in cost-effectiveness analysis. As noted, DALYs combine healthy life years "lost" as a result of disability with healthy life years lost through premature death. Thus the DALY is a negative concept and can be thought of as one lost year of healthy life (Murray & Acharya, 1997). The aim through an intervention is to avert, save or prevent DALYs and to maximize healthy life years (Fox-Rushby & Hanson, 2001). For the formulae used in DALYs and for further reading, see Fox-Rushby & Hanson (2001), Homedes (2000) and Murray & Acharya (1997).

Disability is taken into account by attaching weights to individual illnesses. The weights vary from 0 to 1 according to six disability classes measuring the extent of loss of physical functioning associated with a certain condition: 0 is perfect health (so no deduction of DALYs is made) and 1 is death. For example, class 4 (which includes some cases of dementia and some of blindness) has a disability weight of 0.6 (World Bank, 1993). Note that the disability weights do not take account of the different ways in which individual and social resources compensate for the level of disability experienced (Homedes, 2000).

In the standard model, there is an age weight, which is designed to indicate the increase and then decline in activity of people at different ages. Thus, the age weights used in the standard model rise from birth to reach a peak at 25 years and decline slowly thereafter, as shown in Figure 10.1. Age weights were used in the first edition of *Disease control priorities in developing countries* (Jamison et al. 2006a) but the use of these weights has been controversial. It is argued that DALYs (in the standard model) discriminate against the elderly (Homedes, 2000). In the second edition, the age weights were dropped (Jamison et al., 2006b). As a result, the estimates of the number of DALYs averted and of cost–effectiveness ratios in the second edition differ from those previously published Jamison et al. (2006a).

A discount rate of 3% per year is used in the estimation of a DALY, so that with age weights, the death of a female in infancy corresponds to about 32.5 DALYs lost, the death of a female at age 30 years means the loss of 29 DALYs, and the death of a female at age 60 years represents 12 lost DALYs. Thus, including discounting and age weights, a disease burden of 3300 DALYs in a population would be the equivalent of about 100 infant deaths.

In the standard burden-of-disease calculation, the number of years lost through premature mortality was calculated on the basis of Japanese life expectancies, namely 82.5 years for women and 80 years for men, because life expectancy in Japan is the highest in the world (Homedes, 2000). In the second edition of *Disease control priorities in developing countries*, there was a shift to regional life expectancies, i.e. life expectancies were adjusted to take account of the

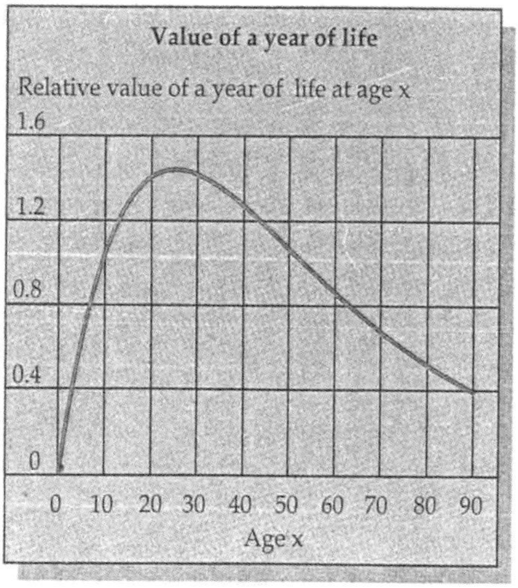

Figure 10.1 Age weights in the 'standard' DALY model. *Source*: World Bank 1993.

realities of specific regions. This had the effect of reducing the cost–effectiveness of interventions in regions with lower life expectancies. As Jamison et al. (2006a, page 279) put it: "For example, averting an infant death in sub-Saharan Africa will save, on average, 44 to 49 undiscounted life years and should not be credited with saving 80 or more".

Major changes were made to the way DALYs are calculated between the first and second edition of *Disease control priorities in developing countries.* Dropping the age weights increases the number of DALYs averted and therefore will reduce costs per DALY for a specific intervention. In contrast, the use of regional life expectancies reduces the DALYs averted and increases the costs per DALY for an intervention.

Given these changes, it is not surprising that a criticism of DALYs is that the method's complexity makes it unlikely that health-care providers or beneficiaries can be involved in a decision-making process based on DALYs (Homedes, 2000). Worse still, as Fox-Rushby & Hanson (2001) point out, rarely do the authors of papers using DALYs declare the assumptions used. Specifically, rarely do authors using DALYs specify the discount rate, the age weight or the life expectancy used. Rarely is sensitivity analysis used in such papers, but sensitivity analysis is important precisely because a change in the underlying assumptions can have significant repercussions (Fox-Rushby & Hanson, 2001). DALYs do have the advantage, however, of allowing conversion of the benefits of many different interventions into a single standard physical value (numeraire).

The focus on measuring health gains in DALYs raises the question of how to include non-health benefits in the cost–effectiveness analysis. Provided these non-health benefits can be expressed in monetary terms, then a frequently used approach is to deduct the benefits from the costs of the intervention as a "negative" cost. For instance, the value of extra economic activity resulting from time saved in collecting water can be deducted from the cost of an intervention that makes water more accessible. That means that the effectiveness indicator (for example, fewer diarrhoea episodes or DALYs) is unchanged by the inclusion of non-health benefits. One drawback of this approach is the possibility of the non-health benefits being greater than the intervention costs, giving negative net costs. Such a result would need to be carefully interpreted.

WHO PROJECT FOR CHOOSING INTERVENTIONS THAT ARE COST–EFFECTIVE

The WHO project on Choosing Interventions that are Cost-Effective (WHO-CHOICE) is an initiative dating back to the late 1990s (Murray et al., 2000). As part of WHO-CHOICE, WHO promoted a generalized cost–effectiveness analysis. In this context generalized means that the outcomes of cost–effectiveness analyses of a wide range of interventions are averaged. Murray et al. (2000) claim that: "such general perceptions of relative cost-effectiveness, which do not pertain to any specific decision-maker, can be a useful reference point for evaluating the directions for enhancing allocative efficiency in a variety of settings".

WHO-CHOICE therefore aims at standardizing and generalizing cost–effectiveness analysis to inform health policy debates. Such generalized cost–effectiveness analysis can provide a quick and cheap reference point for decision-makers. As Murray et al. (2000) put it: "We believe that the more general use of CEA [cost–effectiveness analysis], to inform sectoral debates on resource allocation, is where CEA can make the greatest contribution to health policy formulation".

Murray et al. (2000) point out that cost–effectiveness analysis can develop in two different directions: being increasingly geared to specific situations; or moving towards more generalized assessments. Murray et al. (2000) argue that the time and cost involved, as well as the inherent complexity of taking into account local constraints, will inevitably limit the practical use of cost–effectiveness analysis in specific situations. Instead, the direction that they advocated was:

> *"to focus on the general assessment of the costs and health benefits of different interventions in the absence of various highly variable local decision constraints. A general league table of the cost–effectiveness of interventions for a group of populations with comparable health systems and epidemiological profiles can make the most powerful component of CEA [cost–effectiveness analysis] readily available to inform health policy debates".*

Following the WHO-CHOICE initiative, a manual entitled *WHO guide to cost–effectiveness analysis* was published in 2003 (Tan-Torres Edejer et al., 2003) with the first of the chapters being headed "What is generalized cost–effectiveness analysis?". That chapter draws heavily on the paper by Murray et al. (2000).

In summary, the generalized cost–effectiveness analysis of WHO-CHOICE is a compromise between two extremes. The first extreme is a single "global average" estimate for an intervention's cost–effectiveness, which is likely to be of little value because the costs and effectiveness of any health intervention will vary from one setting to the next. They will vary because epidemiology, baseline levels of infrastructure, the history of disease control, and health promotion vary across countries. However, the other extreme – the ideal of specific estimates for each intervention in every country – is not achievable in the short run. As a compromise, WHO-CHOICE is producing databases reporting the costs and effectiveness of interventions for 14 subregions that have been grouped together on the basis of epidemiology, infrastructure and economic situation.[2]

The next section looks at the application of generalized cost–effectiveness analysis to water improvements.

GENERALIZED COST–EFFECTIVENESS ANALYSIS AS APPLIED TO WATER INTERVENTIONS

The following discussion of generalized cost–effectiveness analysis, as applied to water interventions, is based on a paper by Clasen et al. (2007). While the results cited here are of interest, the main purpose of this exposition is to show cost–effectiveness analysis as applied to water interventions. Clasen et al. (2007) use a variant of generalized cost–effectiveness analysis and compared different

---

[2] See www.who.int/choice/description/need_databases/en.index/html.

interventions on a sector-wide basis for a group of populations with comparable health systems and epidemiological profiles. The two WHO epidemiological subregions that Clasen et al. (2007) look at are Afr-E (sub-Saharan African countries with very high adult and child mortality) and Sear-D (South East Asian countries with high adult and child mortality).

There were five steps in the analysis (see Clasen et al., 2007):

- *Defining the baseline state and the interventions to be investigated.* Clasen et al. (2007) looked at five interventions. One intervention consisted of improving the water supply at source, by using dug wells, boreholes or standposts. The other four interventions consisted of treating the water at the household level (through chlorination, filtration, solar disinfection, and combined flocculation and disinfection). These latter interventions are referred to here as point-of-use treatments. The baseline was taken as 2002.
- *Estimating the costs associated with the interventions.* The mean annual cost per person was estimated for each intervention, using full economic costing and including all costs regardless of whether they are paid by the government, donors, programme implementer or individuals. For source-based interventions, the costs were derived from region-specific estimates of construction and maintenance costs, with the capital costs being annualized using a 3% per year discount rate. The costs of the point-of-use treatments were collected from programme implementers using a set of worksheets developed in accordance with WHO-CHOICE methods. Cost savings that would accrue to the health sector or households in the form of direct costs averted because of reduced levels of diseases were deducted from the gross costs.
- *Estimating the effect of the interventions in preventing endemic diarrhoea.* The relative risk estimate was in the range of 0.63 to 0.73 for all five interventions except household filtration for which the risk estimate was 0.37. The percentage risk of waterborne diarrhoea remaining after the intervention and relative to the baseline was lowest for filtration (at 37%) and highest for the source intervention (at 73%). Thus, of all five interventions, filtration was the most effective in reducing waterborne diarrhoea, although this does not mean that it was the most *cost-effective*.
- M*odelling the population of the two subregions based on demographic, exposure and risk data.* The populations of the two subregions were modelled using a programme called Pop_Mod, a population-modelling programme developed by WHO.
- *Estimating the DALYs averted by each of the interventions*. The DALYs averted by each of the five interventions were calculated, assuming that

the interventions would be implemented for 10 years. This was done using Pop-Mod, which is run once to get the baseline situation and then run again after feeding in the risk reduction. A comparison of the two situations produces an estimate of the DALYs averted.

The results of the study by Clasen et al. (2007) for the Afr-E subregion are shown in Table 10.5. As can be seen from the table, the most cost-effective of the five interventions on both the gross and net cost basis was chlorination, with solar disinfection running as a close second. The source-based intervention was third in the ranking of cost–effectiveness. Clearly, the two best point-of-use interventions would seem to be highly desirable, both showing *net savings* per DALY averted due to reduced health care expenditures.

Table 10.5 Costs and cost–effectiveness in sub-Saharan African countries with very high adult and child mortality[a]

	Intervention				
	Source-based	**Chlorination**	**Solar disinfection**	**Filtration**	**Flocculation and disinfection**
Best estimate of the gross annual costs per person reached (US$)	1.88	0.66	0.63	*3.03*	4.95
Best estimate of the gross cost per DALY averted (US$)	123	53	61	*142*	472
Ranking	3	1	2	*4*	5
Cost savings per person reached (US$)	1.77	1.45	1.20	*0.74*	1.21
Net cost per DALY averted (US$)	7	–63	–55	*107*	357
Ranking	3	1	2	4	5

[a]WHO sub-region Afr-E.
Source: Clasen et al. (2007).

As Clasen et al. (2007) point out: "under definitions established by the WHO Commission on Macroeconomics and Health, interventions are deemed 'cost-effective' or 'highly cost-effective' for a given country if results show that they avert one DALY for less than three times or one times the per capita national gross product, respectively". Clasen et al. (2007) noted that the average GDP per capita in the Afr-E subregion was US$ 369 in the year covered by the study. That being the case, all of the interventions were highly cost–effective on the basis of their net cost per DALY averted. Even on the basis of gross costs, the two best point-of-use interventions are extremely desirable.

This, however, is not the end of the story, because the results of this study are very different from the results of a study of home-based chlorination and safe water storage among HIV-affected households in rural Uganda. For Uganda, Shrestha et al. (2006) estimated a net cost per DALY averted of US$ 1252; this cost is very large compared to the negative costs for chlorination and solar disinfection shown in Table 10.5.

In the study by Shrestha et al. (2006), the cost per DALY averted was very high both because of high costs and low DALYs averted. Cost estimates for the intervention were high in Uganda because of the provision of a safe water storage vessel along with the costs of the chlorine, and other costs associated with the specific context in which the intervention was delivered. The number of DALYs averted were low because of the quality of existing health services – described by Shrestha et al. as: "the aggressive diagnosis and treatment of diarrhoea that was delivered in the field".

Clearly, the contrast between the study by Clasen et al. (2007) and the study by Shrestha et al. (2006) is significant. Some of the results from various studies are compared in Chapter 12, which also looks at the limitations of cost–effectiveness analysis in the water sector. For, as Clasen et al. (2007) admit: "As a cost-effectiveness rather than cost–benefit analysis, this study also omits the economic value of other benefits (including time savings) which have been shown to ensue from improvements in water supplies".

REFERENCES

Clasen T et al. (2007). Cost–effectiveness of water quality interventions for preventing diarrhoeal disease in developing countries. *Journal of Water and Health*, **5**: 599–608.

Fox-Rushby JA, Hanson K (2001). Calculating and presenting disability adjusted life years (DALYs) in cost–effectiveness analysis. *Health Policy and Planning*, **16**: 326–331.

Gittinger JP ed. (1973). *Compounding and discounting tables for project evaluation.* Baltimore, Johns Hopkins University.

Homedes N (2000). *The disability-adjusted life year (DALY) definition, measurement and potential use.* Washington, DC, World Bank (Human Capital Development and

Operations Policy Working Paper) http://www.worldbank.org/html/extdr/hnp/hddflash/workp/wp_00068.html, accessed 15 June 2007.

Hutton G, Haller L (2004). *Evaluation of the costs and benefits of water and sanitation improvements at the global level.* Geneva, World Health Organization. (For a summary of the report, see http://www.who.int/water_sanitation_health/wsh0404summary/en/, accessed 15 June 2007).

Hutton G, Haller L, Bartram J (2006). *Economic and health effects of increasing coverage of low cost water and sanitation interventions.* Occasional Paper for the 2006 Human Development Report.

Jamison et al. (2006a). *Disease control priorities in developing countries.* Washington, DC, World Bank.

Jamison et al. (2006b). *Priorities in health.* Washington, DC, World Bank.

Murray C, Acharya A (1997). Understanding DALYs. *Journal of Health Economics*, **16**: 703–730.

Murray C et al. (2000). Development of WHO guidelines on generalized cost–effectiveness analysis. *Health Economics*, **9**: 235–251.

Shrestha R et al. (2006). Cost–effectiveness of home-based chlorination and safe water storage in reducing diarrhea among HIV-affected households in rural Uganda. *American Journal of Tropical Medicine and Hygiene*, **74**: 884–890.

Tan-Torres Edejer et al. (2003). *WHO guide to cost–effectiveness analysis.* Geneva, World Health Organization.

World Bank (1993). *World development report. Investing in health.* Washington DC, World Bank.

11

Social cost–benefit analysis – principles

John Cameron

The economic assessment of drinking-water interventions – especially small-scale interventions – is challenging because of the complexity of the information needed to assess all direct and indirect outcomes. Meeting these challenges requires the following steps to be taken:

- *Combine information on physical and socioeconomic systems.* Economic assessment of water and sanitation interventions is expected to cover changes in the physical environment (such as water contamination and environmental pollution) and changes in livelihoods. A systematic framework is needed to keep the analysis manageable.
- *Model causality and linking concepts and variables.* Both physical and human processes are complex in themselves, and even more complex in

© 2011 World Health Organization (WHO). *Valuing Water, Valuing Livelihoods.* Edited by John Cameron, Paul Hunter, Paul Jagals and Katherine Pond. Published by IWA Publishing, London, UK.

combination. Economic assessment also bears responsibility for attributing causality, from the planned activities of a development agency to the impact on human and environmental well-being.

- *Identify observable and measurable indicators.* Logical frameworks have been invaluable for economic assessments in making sure that stakeholders agree on observable indicators (Akroyd, 1999).
- *Cope with data gaps and inaccuracy.* Identifying observable indicators does not guarantee they can be measured accurately. In addition, there may be gaps because baseline data may never have been collected and there may be ethical or political reasons for not identifying or observing control groups. Economic assessment frameworks need to provide explicit room for incorporating such concerns.
- *Weight indicators to form composite indices.* Evaluative comparisons also frequently require aggregating indicators into a single number index. This requires weighting of indicators, de facto a form of relative valuing or pricing.
- *Incorporate time to achieve sustainability goals and uncertainty about achieving the goals.* Sustainability is a keyword in wider development thinking as well as in economic assessment. Sustainability involves explicit consideration of long-term processes. Economic assessments clearly cannot be delayed indefinitely to assess impact and sustainability. Hence, assessment frameworks need to be able to incorporate long-term processes and the associated inevitable uncertainty.

The argument here is that an up-to-date social cost–benefit analysis can help meet all these challenges in appraising and evaluating drinking-water interventions.

SOCIAL COST-BENEFIT ANALYSIS – BACKGROUND

The technique of social cost–benefit analysis was originally developed in the 1960s in response to continuing demands on the State to build basic infrastructure. The technique was prompted by growing confidence in a mixed economy with associated widespread market prices, innovations in electronic data processing capacity, and shortage of investable savings and international purchasing power. In the late 1960s, Little & Mirrlees and UNIDO developed social cost–benefit analysis techniques that gave answers to a number of technical questions in pricing costs and benefits (Little & Mirrlees, 1974 – originally published in 1968). This gave economists the apparent power to make a comparative appraisal of any developmental activities against an international standard in terms of their net benefits to the global human condition. This framework included:

- chains linking any developmental activities to final outcomes;
- a numeraire (a common measure of value, e.g. South African rand at 2008 purchasing power), to give an international standard for comparison;
- relative valuation of activities in terms of socially appropriate shadow prices;
- valuation of time through discounting.

An appraisal or evaluation decision then could be made by ranking activities using net present values or benefit/cost ratios or internal rates of return. The framework also gave systematic insights into choice of techniques and the assignment of distributional weights (Mosley, 2001).

The basic social cost–benefit analysis model builds on standard commercial, financial cost–benefit analysis. Financial cost–benefit analysis is what a commercial enterprise would use to appraise or evaluate an investment activity. The model for financial cost–benefit analysis assumes that the enterprise accepts market prices (including interest on borrowing), pays the taxes it cannot avoid (or evade) and welcomes any subsidies. The model also assumes that if the enterprise can displace or externalize costs onto other economic agents (producers, consumers, government, neighbours, the human species), it will. The end result is simply financial profitability for the enterprise as a single institution. Neoclassical economists would claim that this is necessary and sufficient for appraising activities, and that free markets will deliver the best of all possible economic worlds as part of a wider neoliberal developmental agenda (Lipton, 1987).

But social cost–benefit analysis claims the right for an analyst to modify the prices used in the commercial accounts (Al-Tony & Lashine, 2000). This modification is claimed to be valid if and when competitive market forces are not operating as assumed by neoclassical economics or the distribution of wealth is not considered to be just. Criteria to evaluate markets using structure, conducts and performance analysis are presented in Box 11.1.

Social cost–benefit analysis claims to capture market "failures" such as:

- *Absent markets*. Many environmental goods do not have markets. They may be treated as common or pooled property, to which people have access through rights, or which people simply appropriate as they wish. Where assets are being depleted, such as non-recharging fossil water

Box 11.1 Evaluating markets using structure, conduct and performance analysis

What is a perfectly competitive market?

A perfectly competitive market needs:

- a well-specified good or service;
- independent demanders of the good or service, with well-defined tastes;
- independent suppliers of the good or service, with a well-defined technology;
- an institutional framework in which demanders and suppliers meet as well-informed equals to engage in voluntary contracts.

A perfectly competitive market theoretically results in:

- an equilibrium price;
- stability;
- efficiency;
- equity.

A system of perfectly competitive markets theoretically results in:

- general equilibrium and, perhaps, security;
- Pareto superiority and, perhaps, harmony.

supplies or water sources depleted at a faster rate than they can re-charge in drinking-water systems, then social cost–benefit analysis can give a price to that resource by valuing this non-sustainability in terms of future costs of supplying an alternative supply.

- *Externalities.* People's lives may be affected by activities in ways that do not enter into any commercial accounts. People may experience monetary or non-monetary costs and benefits as a result of such activities – air and water pollution are obvious examples.
- *Public goods.* An activity may allow people to get benefits without paying for them individually or preventing others from consuming (e.g. information on a poster at a communal tap giving health information on cleaning containers). People can "free-ride" once the poster is up, because the supplier cannot be sure how much benefit anyone is receiving individually.
- *Imperfect competition.* An imperfect market will allow some economic agents to use power to become price setters in their own interest. Monopoly producers can set prices above the social optimum, while

monopsonistic purchasers[1] can set prices below the social optimum. Control of a spring was used as an early example in economics of the effect of monopoly on pricing.

- *Civil society institutional conventions*. Social conventions may not permit the full operation of market forces. The buying and selling of some goods and services will not be allowed. For other goods and services, prices may only operate within a socially restricted range – common property or pooled rights over forest or water usually have such characteristics.
- *Government regulatory and fiscal actions*. Governments intervene to affect many markets through regulations, taxes and subsidies at regional, national and international levels. Varying taxes and subsidies on agricultural products has implications for imputed values of sanitation and water.

In social cost–benefit analysis, all these forms of market failure could justify modifying observed prices to so-called "shadow prices". A shadow price may be higher or lower than observed prices depending on the specific nature of the market failure. Social cost–benefit analysis involves establishing the value of an activity from the public perspective; at its most ambitious this is a global perspective.

Social cost–benefit analysis always involves judgements on accuracy of data, and the interpretation of data as shadow prices has to take into account the risks and uncertainty surrounding the future. Therefore, sensitivity tests are always needed. The need for sensitivity tests somewhat undermines the claim that social cost–benefit analysis ranks developmental activities on purely technical criteria. But social cost–benefit analysis (through sensitivity testing) does allow healthy deliberation over variables that may crucially influence project performance. The need for sensitivity tests was seen as a weakness in the 1970s and contributed to social cost–benefit analysis being perceived as "smuggling" political judgements into technical assessments. This explicit concern with data inaccuracy and conceptual interpretation is now seen as strength rather than a weakness.

Much of the original work on social cost–benefit analysis focused on government interventions affecting markets through regulations, taxes and subsidies at regional, national and international levels. But economic strategy over the past 30 years across the globe has substantially reduced governmental interventions, removed regulations and reduced variations in taxation and subsidy rates. Generally, confidence in open market forces was high among leading development funders (external support agencies) in the 1980s and the

[1] A monopsonist has market power, because he or she can affect the market price of the purchased good by varying the quantity bought.

influence of social cost–benefit analysis on resource allocation declined. From the early 1980s, neoclassical economics was predominant, claiming that market forces would work directly and indirectly to prevent economic and environmental crises. In the shorter term, markets would ration non-renewable and difficult to renew resources by price rises. In the longer term, profits-induced technological change would prevent environmental melt-down. But such reasoning did not engage with the realities of non-substitutability and irreversibility in many physical environmental processes.

Today, social cost–benefit analysis is being increasingly used as a technique for including environmental factors in projects (Pearce, 1993; Vanclay & Bronstein, 1995; Quah & Tan, 1999; Wattage et al., 2000; Crookes & de Wit, 2002). Most funding agencies wish to incorporate environmental concerns in sanitation and water conservation projects alongside socioeconomic factors. Social cost–benefit analysis can help achieve this aim.

Where a project is causing negative environmental effects, then there are techniques for making explicit calculations of the social costs of the damage:

- If the damage is reversible, then costs for reinstating the natural environment of the project to the pre-project condition should be included in the social cost–benefit analysis, even if this reinstatement is unlikely to occur.
- If the project generates waste or involves resettlement, then costs for the environmentally responsible (including responsibility for health) disposal of that waste or resettlement should be included in the social cost–benefit analysis.
- If the damage to landscape quality is irreversible, then costs of a compensating environmental improvement, not necessarily in the project area, should be included in the social cost–benefit analysis.

The precautionary principle states that if a project has great environmental uncertainties, then, given the complexity of eco-systems, the project should be postponed and only implemented if and when we possess sufficient knowledge of the eco-system to act with reasonable certainty that the environmental risk is acceptable. The precautionary principle is very risk averse, being concerned that unforeseen spread effects may cause disastrous damage. Recognizing that eco-systems are complex, the precautionary principle states that, if the environmental uncertainties are great, then a project should be postponed. As a corollary to the precautionary principle, a project can go ahead if and only if we possess sufficient knowledge of the eco-system to be reasonably certain that the project will not create unacceptable environmental damage. Social cost–benefit analysts may be in the front line of identifying the need to exercise the precautionary principle, because they must identify risks and uncertainties as part of the social cost–benefit analysis procedure.

High discount rates tend to work against environmental responsibility. For example, reinstatement costs at the end of a 20-year project discounted at 12% are worth only a tenth of what they would have been at the start of the project. A 40-year project reduces their value to a hundredth. But social cost–benefit analysts have a responsibility to build environmental costs into assessments, even if these costs may never be paid – the social cost–benefit analyst's task is to assess the full social benefits and costs of an activity, not to compromise with de facto implementation.

In the mid-1970s, many social cost–benefit analysts thought they could introduce distributional concerns into social cost–benefit analysis by identifying the costs and benefits associated with particular social groups and allocating them different weights. The weights were always a matter of judgement – and in that sense political. This was seen as a weakness at that time, and contributed to the side-lining of social cost–benefit analysis.

In the 1990s, governments were making political statements about the importance of reducing poverty and promoting gender equality. A social cost–benefit analysis could support this strategy by applying weights to costs and benefits accruing to women and people judged to be in poverty. It is now much more acceptable not to have technical closure in project decision-making, but rather to present decision-makers with a range of choices, including differing distributional weightings for groups of people with differing socioeconomic characteristics.

Conditions for a perfectly competitive market

Social cost–benefit analysis seeks in principle to value all goods and services as if they were traded in perfectly competitive markets – such market prices are seen as reflecting social valuations. A perfectly competitive market in economics is not directly observable. An important criterion for a perfectly competitive market is that interaction between demand and supply determines the price, not any specific producer or consumer. In other words, in a perfectly competitive market, everyone is a price-taker rather than a price-maker.

Putting monetized values on changes induced by an intervention requires judgements and some knowledge of economic analysis (beyond conventional financial accounting). The analysis involves estimating what markets would do if they were operating freely and universally in conditions approaching perfect competition, in other words if goods and services were being bought and sold until the point where price equates supply and demand, with no market imperfections.

The following conditions are likely to improve the competitiveness of a market:

- active information flows with widespread access;
- low costs of entry into and exit out of a market;
- flexible technology, in terms of variable scales of production, to produce goods and services of comparable quality.

But, in practice, it is difficult to determine how competitive a market is simply by observing the market. Some indication can, however, be gained by asking the following questions relating to the structure, conduct and performance of the market.

Structure

- How many suppliers and demanders are in the market?
- Is an effective framework for legal redress in place?
- Are there barriers to information flows in the institutional framework?

Conduct

- Do transactions occur frequently?
- Are contracts transparent and fair to supplier and consumer?
- Do suppliers and demanders frequently enter into and exit from the market?

Performance

- Is the spatial pattern of prices closely related to transport costs? For example, does the price in area A equal the price in area B plus the transport cost from B to A?
- Do prices move in an economically rational fashion? For example, are prices in step with seasonality in production?
- Are price fluctuations swiftly damped after an external shock?
- Are there signs of large, permanent profits in the system, in terms of growing economic inequality?

If all these questions can be answered positively then the market approaches perfect competition – if not then the analyst needs to reflect on the effects of the negative answer on the observed price. The more nearly the market reflects perfect competitiveness in terms of structure, conduct and performance, the more confidence we can have in the current market price as an indicator of the social worth of a good or service.

Social cost–benefit analysis also requires reflection on causalities to remove double counting. For example, an improved kitchen-garden produces food of greater value (whether the food is consumed by the household or sold), and the value of the kitchen-garden land increases (whether or not the household has any intention of ever selling the land). But changes in both measures of monetary value are caused by the same gain in social value – one directly as an income flow, the other indirectly as a wealth gain. They cannot both go into the social cost–benefit analysis, but which one is used is a matter of empirical convenience, not analytical rigour.

Social cost–benefit assessments of monetized values can also include explicit judgments about distributional social justice. Combinations of judgements on incomplete or imperfect markets, causalities and social justice give rise to so-called shadow prices, monetized values over which the analyst has made choices. These choices should, of course, be stated explicitly in the documentation that accompanies the social cost–benefit analysis.

All the values of the variables can be represented in annualized time profiles moving through a matrix, as shown in Table 11.1.

Table 11.1 Indicative matrix for an intervention showing differing patterns of movements of variables across time

Variable	**Year 0**	**Year 1**	**Intervening years**	**Year X (end of intervention)**
Variable A value	High	Zero	Periodic maintenance for optimum performance	Estimated value of restoring environment to pre-intervention state
Variable B value	Zero	Positive	Constant	Discounted values beyond year X
Variable C value	Zero	Low	Steadily rising and then falling	Zero

Note: Year 0 is start year of intervention; year X is when the intervention is considered not worth continuing for technological or cost reasons (although some variables may continue to have positive values after year X, for example incomes of people whose lives were saved by the intervention).

The social cost–benefit analysis matrix may well look decades into the future and thus involve considerable risk and uncertainty. The effects of this can be incorporated in the matrix by varying the time profiles of key variables to create sensitivity tests.

WILLINGNESS TO PAY

Given all the complexities in using social cost–benefit analysis, plus the risk that expert judgments are being smuggled in disguised as technical truths and not being subjected to sensitivity tests, it is not surprising that so many evaluators of interventions have emphasised willingness to pay as a valuation technique (Piper & Martin, 1999; Ranasinghe, Bee-Hua & Barathithasan, 1999; Vaughan et al., 2000).

Willingness to pay as a criterion avoids all the complications of listing and valuing by simply asking end-users how much they would be willing to pay for an intervention. This assumes that, as rational people, end-users will factor in all the changes they expect, remove all double-counting, and put values on the net changes, to arrive at a single aggregate monetary value. This then is the maximum price the end-users would be willing to pay for their share of benefits from the intervention.

But can such complete information and benign decision-making be assumed? Will a head of household think disinterestedly about the welfare of all household members, and will neighbours recognize their shared interest in a healthier environment (externalities)? Also, the question needs to have been asked in a way that prevents a free-rider undervaluing (if there is a prospect of really paying) or overvaluing (if paying is unlikely and there is a prospect of not receiving the service) the intervention.

In addition, in social justice terms, how can hypothetical willingness to pay be divorced from de facto ability to pay or lack of effective demand? For example, poorer people may express a lower willingness to pay than richer people, not because they value the intervention less, but because they have a different scale of financial valuation. Also, people may associate the question with possible prices that they will have to pay in the future, and hence they would have a material interest in understating the value of the service. The assumptions surrounding willingness to pay as an estimate of monetary value are as demanding as those required for a full social cost–benefit analysis. The empirical concept of willingness to pay may look simpler, but asking the question – 'What would you be willing to pay for improved access to safer drinking-water?' – in a naive fashion makes interpretation of the answer very difficult.

THE SPECIFICS OF SOCIAL COST-BENEFIT ANALYSIS FOR DRINKING-WATER INTERVENTIONS

There is a wide range of variables that may be relevant to a full social cost–benefit analysis of water and sanitation interventions. Each of these variables may merit its own row in the social cost–benefit analysis matrix. For our purposes here, the data requirements are initially presented as empirical questions.

Although the variables are expressed as means, counts and proportions in the questions below, it is important to remember that, in practice, the most accurate way to estimate these aggregate measures is to ask households about their most recent experiences and then aggregate those experiences. That is, do not ask a household: What is the average annual number of "x" episodes? But instead ask a number of households: When did you last experience "x"? The population mean can then be estimated by taking the average lapse in time since the last episode and converting this into an annual rate.

Also worth noting in terms of data collection, is that many of these variables can be estimated using focus group techniques rather than extensive questionnaire surveys.

With both these practical considerations in mind, we can now identify a list of questions that a social cost–benefit analysis may need to see answered:

- What is the mean annual total cash expenditure, if any, of the household on gaining access to drinking-water and sanitation services before and after the intervention? Payments for public or private sources of water will need to be distinguished, because payments for public sources may well have an element of subsidy that will need to be added to give a shadow price reflecting the full social cost. Imperfections in the private sector market, such as elements of monopoly power, may also be identified as price distorting factors.
- What is the cyclical and seasonal pattern of drinking-water access and use? For instance, how many days a year do households use untreated surface water for drinking? What are the health implications of the intervention in changing this pattern of access and use, in terms of DALYs?
- What is the proportion of diarrhoea episodes prevented per year by the intervention (for example, having groundwater available at neighbourhood taps rather than only untreated surface water)? What is the mean incapacitating length of a diarrhoea episode?
- What is the mean time needed for caring, per diarrhoea episode, including accompanying the ill member of the household to seek treatment?
- What is the proportion of people with diarrhoea who seek curative care? What is the mean financial equivalent cost to the household of consultation and treatment for one diarrhoea episode? In answering these two questions, it may be important to distinguish between different socio-economic groups.

- What is the mean or marginal net cost (after deducting any household user payments to the government health service) of treating a diarrhoea episode?
- What is the mean time saved per household per day in collecting water as a result of the intervention (including travel and waiting time for tap water)? Whose time is being saved (by sex and age)?
- What was the mean annual expenditure of the household on water containers and wheelbarrows for transporting water? Is this likely to change as a result of the intervention?
- What proportion of women in the household suffer from permanent back pain or a prolapsed womb that might be attributed to long-term lifting and carrying heavy water containers? What is the gain in DALYs from the improvement in women's health as a result of the intervention?
- What was the mean household expenditure on soap and detergents in the year prior to the intervention? Is this likely to change as a result of the intervention?
- What was the proportion of households undertaking measures to protect drinking-water quality at the point of use? What is the cost of these measures in terms of equipment, consumables and time, including fuel for boiling water, chlorine and other chemicals for water protection, and water filters?
- What was the mean household time spent in activities to protect household hygiene, including cleaning the kitchen, washing containers, and laundering clothes and bed linen? Is this likely to change as a result of the intervention?
- What proportion of children aged 6 to 14 years have improved school attendance as a result of improved water supply?
- What proportion of households has water meters? What is the unit cost of receiving water through the meter?

In addition, in many schemes, benefits may accrue to non-household users or users outside the immediate catchment area of the intervention. In such circumstances, the following questions are relevant to the social cost–benefit analysis.

- What is the proportion of total water demand from the scheme attributable to commercial users? What is the unit cost that commercial users are paying for this water?
- What is the proportion of total water taken by water distributors transporting water to areas outside the scheme? What is the unit cost that water distributors are paying for this water?

Much of the literature on monitoring and evaluating water and sanitation interventions understandably has focused on physiological health improvements brought about by reduced exposure to pathogens. But these are only one aspect of the potential gains from such interventions. Social cost–benefit analysis aims at building a comprehensive list of all the livelihood effects on all the people affected by the interventions over the whole lifetime of the intervention. For instance, if the intervention has a positive impact on productivity in kitchen gardens, perhaps by providing "grey" water irrigation and good quality compost, then this should be included as a benefit. There may be increased food availability for household consumption or sale. Plus, the food may be nutritionally superior and reinforce the direct health gains from the intervention itself.

As we have seen in Chapter 10, in order to assess the impact of the intervention on human well-being, health improvements can be converted into gains in time (measured, for example, as DALYs) available for pursuing valued activities. Some non-health benefits can also be seen as time gains, such as a reduction in the time required to collect water. All such time gains can be given a value in terms of the most valuable livelihood use of the time released (the opportunity cost), even if that time use is not itself monetized. By making comparisons with broadly equivalent monetized activity, a value can be found. For example, more time spent in cultivation for household consumption may be valued at the local agricultural wage rate for paid labour.

Some benefits may not be reducible to more time available, such as the improvement of a kitchen garden. Thus, time saving is not a universal standard of value (or numeraire). By assigning monetized values to changes induced by an intervention, it is possible not only to include all the changes but also to differentiate between different uses of saved time, reflecting their differing worth to society as a whole, for example time used in socializing in a bar compared with time hoeing a kitchen garden. Such differentiation allows social cost–benefit analysis to provide a more subtle understanding of relative gains from different interventions, compared to valuing only time gains. Table 11.2 indicates the range of possible benefits, linked to potential indicators, that might result from a drinking-water intervention.

In terms of framing the social cost–benefit analysis, the target population will need to be segregated by age and sex. Information on population movements, permanent, cyclical and seasonal, may well be relevant. Given the long time horizon for many water and sanitation interventions, it may be worth investing time and resources in developing a full demographic model showing population change. This model can be brought into interaction with changing exposure rates induced by the water or sanitation intervention. The most dramatic effect

Table 11.2 Forms of possible livelihood benefits resulting from a drinking-water intervention

Benefit variables	Indicators (pre- and post-intervention mean values)
Sickness time saved (economically active adults)	Number of annual ill-health episodes and mean length of episode
Sickness time saved (economically inactive adults)	Number of annual ill-health episodes and mean length of episode
Sickness time saved (children)	Number of annual ill-health episodes and mean length of episode
Benefits from mortalities postponed per household	Mean net income earned per person over additional years
Caring time saved (economically active adults)	Mean length of caring time per ill-health episode
Caring time saved (economically inactive adults)	Mean length of caring time per per ill–health episode
Caring time saved (children)	Mean length of caring time per ill–health episode
Household health-care costs saved	Mean household health-care cost per ill-health episode (including transport)
Government health-care costs saved	Mean cost to government of providing the health care per ill-health episode
Water collection time saved per household	
• economically active adults	Mean hours per day (including waiting time)
• economically inactive adults	Mean hours per day (including waiting time)
• children	Mean hours per day (including waiting time)
Disabilty damage prevented in water collection per household	Mean health-care costs plus loss of income earning capacity
Gains from improved educational performance	Percentage of children improving school attendance, by age
Value added gains from additional irrigated crops	Mean increased value of crops cultivated less all input costs
Environmental gains (What proportion of households considers that the changes in the water and sanitation system have changed the physical environment of the village? What proportion thinks the changes have been for the better? What proportion thinks that the changes have been for the worse?)	Total increased amenity value of land in area

(*Continued*)

Table 11.2 *Continued*

Benefit variables	**Indicators (pre- and post-intervention mean values)**
Social capital benefits (What proportion of households considers that the changes in the water and sanitation system have changed the social atmosphere in the village? What proportion thinks that the changes have been for the better? What proportion thinks that the changes have been for the worse?)	Percentage people stating increased confidence and trust in planning and implementing developmental activities
Proportion of financial benefits devoted to productive investment in equipment or tools	Proportion of additional income from health and non-health benefits invested to increase future income

of interventions on demographic variables is reduction in infant and child mortality, but accurate data on changes are very difficult to collect for relatively small affected populations. Therefore larger scale population data sets enabling comparison of relatively well-provided and less well-provided groups of people, may be needed to get a reliable parameter on mortality rates for the affected population.

Estimating lifetime earnings for those lives saved will require judgments to be made on long-term economic change and will be sensitive to the discounting rate over time, given that much of this gain will be far in the future. Considerations of social justice may be significant here in terms of the value of a human life being seen only as the discounted value of future earnings. Also, the impact of greater child survival on future fertility choices may be a significant externality that needs to be incorporated into the demographic model.

REVIEW OF THE COMPLETE PROCESS FOR SOCIAL COST–BENEFIT ANALYSIS

The process of social cost–benefit analysis can begin by using a modified logical framework to identify the numerous linkages between activities and final impacts, using a brainstorming approach with people who have had direct experience of similar activities in the locations where the development agency is operating. The brainstorming also can use a stakeholder model to identify groups of people likely to be affected, both positively and negatively, by the activities. Risk analysis can be used to identify variables that would be relevant to sensitivity tests.

With the results of these brainstorming activities to hand, a list of all physical inputs, outputs and effects giving rise to costs and benefits can be made in a single column on an Excel spreadsheet. A timescale in terms of years is then put on to the columns of the spreadsheet, starting from the year of construction. Social cost–benefit analysis also requires the choice of a final year.

The final year will be: either when flows of costs and benefits have steadied and discounting reduces present values to insignificant levels, or when a scheme is believed to require such heavy capital expenditure that a substantial new investment activity would be needed beyond normal operation and maintenance activities.

For that final year, decisions have to be made, preferably at the appraisal stage (or, if not, in the impact evaluation), about the socially acceptable environmental status it should obtain (or should have obtained) at the end of project. There may need to be a complete reinstatement of the pre-intervention conditions, or an environmentally compensating activity elsewhere.

Estimates of physical quantities of inputs and outputs are then made and fed into the timescale in the years they affect. Market prices for unit quantities of all inputs and outputs are identified wherever possible from primary and secondary sources. Each of these market prices is scrutinized and discussed to decide whether it should be modified to a socially more appropriate shadow price. The scrutiny focuses on institutional factors that are affecting the observed prices and then modifies these observed prices, in the direction of removing the institutional effects to reveal a shadow price. In some cases, such as taxes and subsidies, the effects can be relatively easily quantified. In others, such as foreign exchange rates, standard formulae often exist at national level. Where there are believed to be private monopolies controlling peak-season transactions, the effects can be quantified by finding low-season transactions outside the monopoly relationship and assessing where a market clearing price might lie, taking the scale of activity into account.

In some cases, missing prices for some inputs and outputs can be derived from related inputs and outputs that have observable prices. Land is an example, even though not readily bought and sold in many rural societies. Changes in land use can be given an imputed price by observing changes in net value of the produce from the land.

The challenges are greater where the physical changes induced by the activities of the development agency have no observable prices, or the linkages are very indirect. Best estimates of values for such changes can be made using secondary material where available or allocating a notional value added. Such estimates would be prime candidates for sensitivity tests.

Once a complete set of shadow prices representing social values and real scarcities has been established, then these are applied to the quantities of inputs and outputs, and a spreadsheet expressed solely in shadow price values is produced. Where particular costs and benefits are seen as accruing heavily to vulnerable or other target groups, weightings could be applied to reflect distributional concerns. All assumptions about the time pattern of physical activities, pricing of resource use and physical effects, and distributional weightings should be fully documented.

The spreadsheet is then expanded by three rows to calculate total costs and total benefits, and net costs/benefits in each year. A standard discounting formula is applied to net costs/benefits in each year to take account of the effects of time. Either net present values can be calculated for a target discount rate, or an internal rate of return (reducing the sum of discounted net costs and benefits to zero) can be calculated as a first complete scenario.

This scenario presents the most likely estimate. It can then be adjusted to best case and worst case scenarios by reviewing the risk analysis and the assumptions made. All risks and assumptions that tended to increase benefits and decrease costs are quantified and used for a best case scenario. To construct a worst case scenario, all risks and assumptions that would increase costs and decrease benefits are quantified.

All three scenarios, with their associated documentation, can then be offered to decision-makers for consideration in the final assessment process. Presenting at least three scenarios is a clear signal to the decision-makers that the social cost–benefit analysis is not merely a technical calculation, but an indicative exercise in which judgement must be exercised by decision-makers.

Social cost–benefit analysis is not presented here as a panacea for appraising or evaluating water interventions. Social cost–benefit analysis can help in making an assessment but cannot determine a decision on the results. The practice of social cost–benefit analysis has matured since the 1970s. In its best practice, social cost–benefit analysis today is explicit and transparent about assumptions and judgements involved. Social cost–benefit analysis presents decision-makers with choices they can make with good deliberative reason, and does not seek technical closure (Morimoto & Hope, 2004).

In this chapter we have attempted to describe social cost–benefit analysis at its most technically ambitious, in terms of a wide-ranging livelihoods framework. We have found no studies of water interventions that meet the high demands of both rigorous principles and empirical range. We hope that this book will encourage more ambitious efforts to apply social cost–benefit analysis to more local drinking-water interventions in the future. Chapter 12 outlines the best examples currently available of using social cost–benefit analysis to assess water interventions at a highly aggregated level.

REFERENCES

Akroyd D (1999). Logical framework approach to project planning, socio-economic analysis and to monitoring and evaluation services: a smallholder rice project. *Impact Assessment and Project Appraisal*, **17**: 54–66.

Al-Tony FES, Lashine A (2000). Cost-benefit analysis of railway electrification: case study for Cairo-Alexandria railway line. *Impact Assessment and Project Appraisal*, 18: 323–333.

Crookes D, de Wit M (2002). Environmental economic valuation and its application in environmental assessment: an evaluation of the status quo with reference to South Africa. *Impact Assessment and Project Appraisal*, **20**: 127–134.

Lipton M (1987). Limits of price policy for agriculture: which way for the World Bank? *Development Policy Review*, **5**: 197–215.

Little I, Mirrlees J (1974). Project appraisal and planning for developing countries. London, Heinemann.

Morimoto R, Hope C (2004). Applying a cost-benefit analysis model to the Three Gorges project in China. *Impact Assessment and Project Appraisal*, **22**: 205–220.

Mosley P (2001). A simple technology for poverty-oriented project assessment. *Impact Assessment and Project Appraisal*, **19**: 53–67.

Pearce DW (1993). Economic values and the natural world. London, Earthscan.

Piper S, Martin W (1999). Assessing the financial and economic feasibility of rural water system improvements. *Impact Assessment and Project Appraisal*, **17**: 171–182.

Quah E, Tan, KC (1999). Pricing a scenic view: the case of Singapore's East Coast Park. *Impact Assessment and Project Appraisal*, **17**: 295–303.

Ranasinghe M, Bee-Hua G, Barathithasan T (1999). Estimating willingness to pay for urban water supply: a comparison of artificial neural networks and multiple regression analysis. *Impact Assessment and Project Appraisal*, **17**: 273–281.

Vanclay F, Bronstein DA, eds (1995). *Environmental and social impact assessment*. Chichester, Wiley.

Vaughan WJ et al. (2000). Uncertainty in cost–benefit analysis based on referendum contingent valuation. *Impact Assessment and Project Appraisal*, **18**: 125–137.

Wattage P et al. (2000). Integrating environmental impact, contingent valuation and cost-benefit analysis: empirical evidence for an alternative perspective. *Impact Assessment and Project Appraisal*, **18**: 5–14.

12

Social cost–benefit analysis – summarizing the available global evidence on drinking-water interventions

Chris Edwards

If we want to combine all types of benefits from drinking-water interventions (health gains, increased income, time saved) on the outcomes side of an economic assessment, then we have to evaluate all the outcomes in monetary terms (see Jamison et al. 2006a). This means using social cost-benefit analysis as discussed in previous Chapters.

The chapter on water in *Disease control priorities in developing countries* (Cairncross & Valdmanis, 2006) points out that most investments in water improvements come from the public sector, but not from health sector budgets. As a result water improvements are often health-related interventions that come without cost to the budget of the health sector. But if they *are* appraised on health grounds alone, then they are likely to be given a much lower ranking than they

© 2011 World Health Organization (WHO). *Valuing Water, Valuing Livelihoods*. Edited by John Cameron, Paul Hunter, Paul Jagals and Katherine Pond. Published by IWA Publishing, London, UK.

deserve compared to other investments, because these calculations omit the valuation of significant time savings in collecting water. Drinking-water interventions are therefore complex in terms of both responsibility for costs and accounting for benefits. As a result, drinking-water interventions, especially small-scale interventions, will tend to be undervalued in terms of both ownership of costs and counting of benefits.

The data presented in Table 12.1 show that the benefit/cost ratios for water and sanitation projects are high in all areas of the world. As Figure 12.1 shows, in this case for sub-Saharan Africa, the vast majority of benefits is derived from time savings.

Table 12.1 Annual cost estimates, total economic benefit estimates and benefit/cost ratio for achieving six water and sanitation coverage scenarios, by world region

World Region	MDG target			Universal access		
	Water	**Sanitation**	**W&S**	**Water**	**Sanitation**	**W&S**
Annual cost estimates (US$ millions)						
Sub-Saharan Africa	479	2 185	2 665	777	3 379	4 156
Arab States	66	188	254	96	492	589
East Asia & Pacific	229	399	628	891	4 576	5 468
South Asia	53	802	856	189	5 033	5 222
Latin America & Caribbean	14	219	233	87	734	821
Eastern Europe & CIS	16	19	35	34	292	326
Non-OECD	858	3 813	4 671	2 075	14 507	16 581
Total economic benefit estimates (US$ millions)						
Sub-Saharan Africa	1 336	14 359	15 292	3 006	21 963	23 566
Arab States	403	1 005	1 375	572	6 230	6 680
East Asia & Pacific	1 593	5 003	6 364	5 883	63 093	66 825
South Asia	186	5 507	5 635	733	34 305	34 706
Latin America & Caribbean	110	8 287	8 352	1 498	28 787	29 801
Eastern Europe & CIS	133	542	671	307	8 711	8 930
Non-OECD	3 762	34 703	37 689	11 999	163 088	170 508
Benefit/cost ratio						
Sub-Saharan Africa	2.8	6.6	5.7	3.9	6.5	5.7
Arab States	6.1	5.3	5.4	5.9	12.7	11.3
East Asia & Pacific	6.9	12.5	10.1	6.6	13.8	12.2
South Asia	3.5	6.9	6.6	3.9	6.8	6.6
Latin America & Caribbean	8.1	37.8	35.9	17.2	39.2	36.3
Eastern Europe & CIS	8.3	27.8	18.9	8.9	29.9	27.4
Non-OECD	4.4	9.1	8.1	5.8	11.2	10.3

Source: Hutton, Haller & Bartram, 2007.

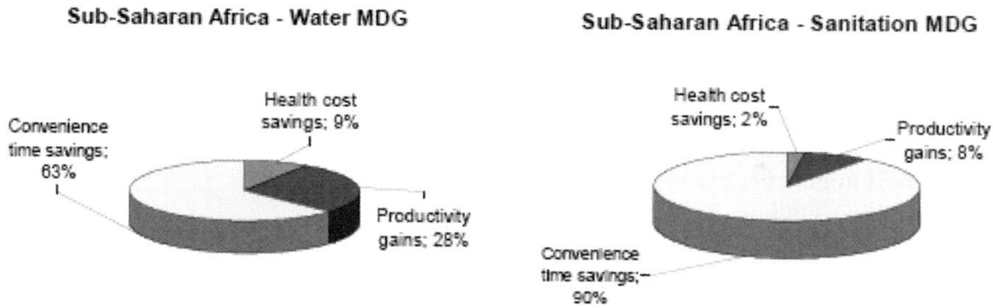

Figure 12.1 Contribution of major benefit categories to total economic benefit in sub-Saharan Africa for meeting the water (left) and sanitation (right) MDG targets.

Having defined the current global situation, identified feasible interventions to improve water, estimated costs and valued benefits, the final stage – discussed in this chapter – is to compare the costs and benefits of actual drinking-water interventions and to give guidance on whether, how and when any of them should be financed.

It is worth noting in Table 12.1 that the benefit/cost ratios for water improvements are high, ranging from 2.8 to 8.3. Yet, as is clear from Figure 12.1, a major part of the benefits (63%) consists of the value of access time saved. In the absence of the savings in time, the benefit/cost ratio for water improvements in sub-Saharan Africa would have been only a little over unity. It is clear that disease reduction can be brought about by point-of-use interventions (such as chlorination) but it is also clear that low-cost investments to bring water supplies closer to the household are likely to generate high economic returns because of savings in access time.

Thus, when looking at the present water situation, it is important to answer the following questions:

- How long does the average household take to collect water per day or per week?
- How long will the average household take to collect water per day or per week after each of the proposed feasible improvements?
- How many households will benefit from the improvements?

To answer these questions, it would be useful to prepare a map of the area showing the locations of households and the existing and possible water facilities.

Some information (taken from other studies) about the time taken to collect water is given by Hutton, Haller & Bartram (2006) and the assumptions made about benefits from reduced access time in their report are summarized in Table 12.2.

Table 12.2 Estimates of the time benefits from reduced access time for water

	Data values	
	Range	**Median**
Time saved from better external access (hours/household per day)	0.25 to 1.0	0.5
Time saved from having water piped to the household (hours/household per day)	1.0 to 2.0	1.5

Source: Hutton, Haller & Bartram (2006).

The savings shown in Table 12.2 are highly spatially generalized. An average saving of 0.5 hours per household per day was assumed for a water improvement. It is clearly important to estimate carefully the time savings per household in a particular local livelihoods context. It is also important to attempt to estimate the benefits from any additional economic activity resulting from improved water facilities. The additional net output which is most likely to arise in a rural area is from increased irrigation of crops made possible by improved water facilities. An estimate of the existing output from small-scale agriculture (including household garden plots) in the area will be required, together with an estimate of the likely additional output from an improved water supply. It may be desirable to undertake agricultural investments jointly with water improvements to maximize the return from additional irrigation. For example it may be desirable to invest in cultivation inputs or extra collection and storage facilities to enhance the additional agricultural output.

The benefits from increased irrigation were not included in the Hutton, Haller & Bartram (2006) study, nor were various other benefits (benefits from more leisure time or activities, from a reduction in vector-borne diseases, from reduced pain and anguish associated with lives lost, or from aesthetic improvements). If there are ways of measuring these and if they are likely to be significant, they should be included.

So far we have talked about benefit/cost ratios but it is worth noting that there are two other ways in which the costs and benefits can be compared. These are net present value and internal rate of return. Annex 1 sets out the three methods and illustrates their use with an invented example of a fictitious investment in the development of a dug well in a stylised village in South Africa.

We now look at how the cost–effectiveness results can be converted into cost–benefit analysis. This means that we need to put a monetary value on a DALY and we discuss that issue in the next section.

THE MONETARY VALUE OF A DALY

In Table 10.5 of Chapter 10, the best estimates of the gross cost per DALY averted for five water interventions ranged from US$ 472 for flocculation and disinfection to US$ 53 for chlorination. To convert such values into benefit/cost ratios, we need to know the monetary value of a DALY.

Unfortunately there is a wide range of estimates for the value of a DALY. Lvovsky et al. (2000, Chapter 4) gives the value of a DALY for six cities in developing countries and countries in transition. The average for the six cities (Bangkok, Krakow, Manila, Mumbai, Santiago and Shanghai) in the study can be estimated at US $11 100. According to the author's estimates for the year 2000, the value of US$ 11 100 is about five times as large as the GDP per capita for the six countries in which these cities are located.

A second estimate is given by the WHO Commission on Macroeconomics and Health (WHO, 2001), which suggested that interventions with an annualized cost of less than three times the GDP per capita are cost-effective. So here we have the value of a DALY given as three times the GDP per capita

By contrast, a much lower valuation of US$ 100 per DALY averted seems to be implied as a threshold value in the 1993 *World development report* (World Bank 1993). However, even if we take the higher figure of US$ 150 quoted from the 1993 *World development report* by Cairncross and Valdmanis (2006), this compares with a per capita GDP for low–income countries given in the same World Bank report of US$ 350 (World Bank 1993) and therefore the threshold (and the implied value of a DALY) of US$ 150 corresponds to only about two fifths of the per capita GDP of the poorest countries. Thus we have monetary values for a DALY ranging from two fifths to five times of GDP per capita.

This wide range may be explained by a number of factors. First, as noted in Chapter 10, different assumptions are used by different authors for the calculation of DALYs. For example, dropping age weights could increase the number of DALYs averted in a given situation. Second, there are two quite different approaches used for valuing a life and putting a value on a DALY. One is the human capital approach and the other is the willingness to pay (or contingent valuation) approach (see Jamison et al. 2006a).

The standard 1993 method for calculating DALYs is based on the human capital approach, and the formula for the DALY would seem to imply that its monetary value is equal to GDP per capita. If we assume a DALY value equal to average GDP per capita, then this converts in 2004 to an average for low-income developing countries of US$ 538 (UNDP 2006). Rounding this figure, a DALY is here assumed to be worth US$ 500 (in 2004 and for low-income developing countries).

WHAT ARE INVESTMENTS IN WATER IMPROVEMENTS WORTH?

While cost-effectiveness analysis may be fine for interventions where benefits can be measured *solely* in terms of the indicator (in this case DALYs averted), it will be of *limited* use in comparing the value of investments in the health sector which produce a mix of benefits, and it is of no value for comparisons with investments in other sectors (such as education or industry).

Furthermore, the cost-effectiveness approach is likely to be confusing for a non-specialist, first because the value of non-health benefits will have to be deducted from the cost in the numerator and second because the meaning of a DALY is by no means obvious.

Converting cost-effectiveness figures into benefit/cost ratios makes it simpler to compare the effects of different water and sanitation improvements. We attempt to make such comparisons below.

Table 12.3 compares the estimates for water improvements given by Hutton, Haller & Bartram (2006), Cairncross & Valdmanis (2006), and Clasen et al. (2007). In fact, Clasen et al. (2007) looked at five types of improvements, including four point-of-use improvements, but Table 12.3 shows only the best of the four point-of-use interventions.

Table 12.3 shows that the benefit/cost ratios are all greater than one. The results of the two cost-effectiveness studies have been converted into benefit/cost ratios by valuing a DALY averted at US$ 500, which is roughly the GDP per capita in developing countries. It is surprising that the benefit/cost ratios are so high in the studies by Cairncross & Valdmanis (2006) and Clasen et al. (2007) because neither of these studies included the benefits from reduced access time in their analysis. These benefits are very important, accounting for almost two thirds of the total in the Hutton, Haller & Bartram (2006) study.

In converting the figures for cost per DALY averted into benefit/cost ratios, the valuation of a DALY is clearly crucial. Table 12.4 shows the benefit/cost ratios for the two cost-effectiveness studies if we value a DALY at 30% of GDP per capita, that is at US$ 150 per capita.

With a DALY valued at US$150, the benefit/cost ratios for the low-cost water improvements fall from 5.3 to 1.6 in the Cairncross & Valdmanis (2006) study and from 5.0 to 2.2 in the Clasen et al. (2007) study. Interestingly, even with this lower valuation of a DALY, both studies show positive benefit/cost ratios and that is without taking account of the benefits from reduced access time.

Hutton, Haller & Bartram (2006) carried out a number of sensitivity analyses. One of these sensitivity tests included the valuation of time at 30% of GDP per capita. This lower valuation was used on the grounds that it is the value that

Table 12.3 The worth of water improvements: a comparison of estimates valuing time savings and DALYs at average GDP per capita (all monetary values in US$; na, not available).

	Study and intervention				
	Hutton, Haller & Bartram (2006)	Cairncross & Valdmanis (2006)		Clasen et al., 2007	
	Water	**Household connections**	**Handpump or standpost**	**Dug well, borehole or stand post**	**Chlorination**
Cost per person reached	2.3	17.5	2.0	1.9	0.7
Cost per DALY averted	na	223	94	123	53
Benefit of a DALY averted	na	500	500	500	500
Value of deaths saved	1.3	na	na	na	na
Health costs saved	1.1	na	na	116	116
Access time saved	4.1	na	na	na	na
Total benefits	6.5	na	na	616	616
Benefit/cost ratio	2.8	2.2	5.3	5.0	11.6

Table 12.4 The worth of water improvements: comparison of estimates (in US$) valuing DALYs at 30% of GDP per capita; na, not available

	Study and intervention			
	Cairncross and Valdmanis, 2006		Clasen et al., 2007	
	House connections	**Handpump or standpost**	**Dug well, borehole and standpost**	**Chlorination**
Cost per person reached	17.5	2.0	1.9	0.7
Cost per DALY averted	223	94	123	53
Benefit of a DALY averted	150	150	150	150
Health costs saved	na	na	116	116
Access time saved	na	na	Na	na
Total benefits	None		266	266
Benefit/cost ratio	0.7	1.6	2.2	5.0

people put on their time according to a study by the International Monetary Funds. At this valuation, the benefit/cost ratio for water improvements for sub-Saharan Africa fell from 2.8 to 1.1 (see Hutton, Haller & Bartram 2006). The valuation problem is discussed in Annex 2.

It is interesting to note that the cost-effectiveness studies of Cairncross & Valdmanis (2006) and of Clasen et al. (2007) seem to show much higher benefit/cost ratios than those estimated by Hutton, Haller & Bartram (2006). The figures for the cost per person reached are similar in all three studies. The differences seem to lie in very different values of benefits. These differences need to be examined carefully in the context of the specific interventions.

All three studies, however, show that low-cost water improvements seem to provide a good surplus of benefits over costs, suggesting that social cost-benefit analyses may well yield positive results in a wide range of contexts[1].

THE WAY FORWARD: POLICY, PACKAGES AND SEQUENCING

Social cost-benefit analysis allows cross-sectoral comparisons, and the aggregate results on existing data do look promising in terms of rates of return. Table 12.3 shows that the benefit/cost ratios for low-cost water and sanitation improvements are positive when a DALY and time saved are valued at per capita GDP. The benefit/cost ratios remain positive even when the valuations are at 30% of per capita GDP.

While more analysis needs to be done in terms of specific interventions, point-of-use treatments do seem to have a better health effect than providing public standpipes. Walsh & Warren (1979) argued that: "… public standpipes … are not highly effective in reducing morbidity and mortality from water-related diseases. It is well-documented that connections inside the house are necessary to encourage the hygienic use of water". This still seems to be the case. Cairncross & Valdmanis (2006) state that "Providing a public water point appears to have little effect on health, even where the water provided is of good quality and replaces a traditional source that was heavily contaminated with faecal material. By contrast, moving the same tap from the street corner to the yard produces a substantial reduction in diarrhoeal morbidity", the reason being

[1] Whittington & Hanemann (2006) reached a very different conclusion: "there is little evidence to suggest that the current monthly benefits of improved water and sanitation services exceed the monthly cost". But they studied the costs and benefits of combined water and sanitation systems in urban areas for which the total cost per person per year was US$ 40. Furthermore, their method of estimating benefits differed from that used in the studies discussed in this Chapter. They placed a heavy emphasis on willingness-to-pay and 'coping' costs.

that "domestic hygiene … is the principal determinant of endemic diarrhoeal disease rates and not drinking-water quality".

Not only does providing a house or yard connection produce markedly better health results than a public tap because it is associated with better hygiene, but as Cairncross & Valdmanis (2006) also point out "collecting revenue from households with private connections is far simpler than collecting it from public taps". Unfortunately, water connections into the house or yard are likely to give benefit/cost ratios of less than one, particularly in dispersed rural villages. Clearly this option – water connections to the house or yard – is one that needs to be subjected to a cost-benefit analysis in a particular situation.

Judging from the studies looked at here it seems, however, that point-of-use treatments (particularly chlorination) would give highly cost-effective health results, while improved access to safe water sources generates valuable savings in time. A sensible investment strategy for water improvements would seem to be to combine point-of-use treatments (particularly chlorination) with closer water supplies. The two types of improvement are not mutually exclusive. Point-of-use treatment seems to be justified (in terms of giving a benefit/cost ratio much greater than one) on the grounds of disease reduction. Improving water supply seems to be justified on the grounds of reduced access time and increased irrigation for agricultural production.

Such an analysis opens the way to the sequencing and packaging of improvements. Social cost-benefit analysis is able to help in this assessment, particularly if the costs and benefits are set out on a year-by-year basis using the net present value approach (as illustrated in Annexes 1 and 2). If investing in the maintenance of an existing system is considered to be potentially superior to investment in a new scheme, this can be analysed by entering the costs and benefits of the schemes on a year-by-year basis, discounting to a net present value and comparing the results.

When the analysis has been completed, we are still left, however, with the problem of financing and implementation. How should or can the improvement be financed? Are the beneficiaries likely to be willing or able to finance the water improvements? Can microcredit schemes be set up to facilitate the financing of the intervention by the potential beneficiaries? And if the beneficiaries cannot finance the intervention, can local authorities finance and maintain the improved supply of drinking-water? These are some of the questions that remain.

REFERENCES

Cairncross S, Valdmanis V (2006). Water Supply, Sanitation and Hygiene Promotion. Ch.41 in Jamison et al. (eds.) *Disease Control Priorities in Developing Countries*. Washington, DC, World Bank. Available at: www.dcp2.org/pubs/DCP

Clasen T, et al. (2007). Cost-effectiveness of water quality interventions for preventing diarrhoeal disease in developing countries. *Journal of Water and Health*, 5: 599–608.

Haller L, Hutton G Bartram J (2007). Estimating the costs and health benefits of water and sanitation improvements at global level. *Journal of Water and Health* 5: 467–480.

Hutton G (2000). *Considerations in evaluating the cost-effectiveness of environmental health interventions*. Geneva, World Health Organization (WHO/SDE/WSH/00.10).

Hutton G & Haller L (2004). *Evaluation of the costs and benefits of water and sanitation improvements at the global level*, (WHO/SDE/WSH/04.04).

Hutton G, Haller L, Bartram J (2006). *Economic and health effects of increasing coverage of low-cost water and sanitation interventions.* Occasional Paper for the 2006 UNDP Human Development Report.

Hutton G, Haller L, Bartram J (2007). *Economic and health effects of increasing coverage of low-cost water and sanitation interventions to countries off-track to meet MDG target 10.* Document WHO/SDE/WSH/07/05. Geneva, World Health Organization.

Jamison et al. (2006a). *Priorities in Health*, Washington DC, World Bank.

Jamison et al. (2006b). *Disease control priorities in developing countries*, Washington DC, World Bank.

Lvovsky K et al. (2000). *Environmental costs of fossil fuels: a rapid assessment method with application to six cities.* Washington DC, World Bank Environment Department. Pollution Management Series.

Walsh JA, Warren KS (1979). Selective primary health care: an interim strategy for disease control in developing countries. *New England Journal of Medicine* 301: 967–974.

Whittington D, Hanemann W (2006). *The economic costs and benefits of investments in municipal water and sanitation infrastructure; A global perspective"*, Berkeley, University of California, Department of Agricultural and Resource Economics.

UNDP (2006). *Human Development Report.* New York, United Nations Development Programme.

World Bank (1993). *World development report, investing in health.* Washington DC, World Bank.

WHO (2001). *Report of the Commission on Macro-economics and Health.* Geneva, World Health Organization.

ANNEX 1. EXAMPLE OF THE COSTS AND BENEFITS OF A DUG WELL

This annex to Chapter 12 presents an (invented) example of an improved water supply in the form of a dug well and shows how it can be analysed in three different ways: benefit/cost ratios; net present value; and internal rate of return. In the example, the total population of the village in 2005 is 600 (120 households), of whom 100 people (20 Households) gain access to an improved water source (a dug well).

BENEFIT/COST RATIO

Table A1.1 analyses the imaginary project in terms of a benefit/cost ratio, the same method as that used by Hutton, Haller & Bartram (2006).

Table A1.1 Costs and benefits of a dug well in a village in South Africa

Costs	Total (in rand)	per person reached	
		Rand	US$
Investment costs[a]	36 000	360	50
Annualised investment costs[b]	4 200	42	5.9
Annual recurrent costs (at 10% of annualised capital costs)	400	4	0.6
Total annual costs	4 600	46	6
Annual benefits			
Health and health Costs	4 600	46	6.4
Access time saved[c]	9 600	96	13.4
Total annual benefits	14 200	142	19.8
Benefit/cost ratio	3	3	3

Notes:
[a] Life of investment 10 years; capital recovery factor at 3% per annum 0.117.
[b] Health costs saved as a result of the improved water supply are assumed to consist of reductions in health system costs, reductions in patient costs, time saved from less illness and the value of reductions in deaths. These are assumed to total about the same as the annual costs.
[c] Access time saved is assumed to be 20 minutes per household per day valued at the GNP per capita in South Africa (US$ 4900 in 2006).

The benefit/cost ratio for this imaginary project is estimated as 3. Here time savings have been valued at the GNP per capita of South Africa. In 2004, however, the

South-African per capita GNP was more than six times as high than that of sub-Saharan Africa as a whole. Great care needs to be taken in assigning a value to the time, because the worth of a water improvement is likely to be highly sensitive to the valuation of time.

If the benefit/cost ratio is above 1, then the investment is worth undertaking, assuming that there is no mutually exclusive, alternative investment promising a higher ratio, and assuming that the interest rate of 3% per annum is a good reflection of the cost of capital.

The disadvantage of the benefit/cost ratio is that the investment costs have to be converted into their annual equivalent. This approach is somewhat less transparent than laying out the costs and benefits on a year-by-year basis which is done for the next measure described, the net present value.

THE NET PRESENT VALUE

In Table A1.2 on the next page, the costs and benefits of the assumed project are shown on a year-by-year basis over ten years which is here assumed to be the effective physical life of the dug well.

The sum of the costs and benefits in each year is usually called the cash flow. If we add these over the life of the project (in this case, ten years), we can see whether we have a surplus of benefits over costs.

From Table A1.2, we can see that we do. The surplus of benefits over costs is rand 73 000 and the project looks very attractive. However this is not the end of the story. The analysis needs to go beyond this by discounting the costs and benefits. In Table A1.2 the discount rate used is 3% per annum.

We can see from the final column of Table A1.2 that the net present value (at a discount rate of 3% per annum) of the project is rand 69 000. This means that the project could pay an interest rate of 3% per annum on the investment and end up with a positive net present value at the end of ten years worth rand 69 000. Indeed the project is so healthy that even if we charge an interest rate of as much as 20% per annum, we still end up with a positive net present value as shown in Table A1.3 worth rand 13 000.

THE INTERNAL RATE OF RETURN

The third cost-benefit measure that we look at here is the internal rate of return (IRR). This is the rate of interest that the project could pay over the ten years of life of the project and still break even – that is end up with a net present value of zero. In this case, the IRR can be estimated to be about 30% per annum, since at that discount rate, the net present value is about zero, as shown in Table A1.4. The attraction of

Table A1.2 The present values of costs and benefits of an imaginary project at a discount rate of 3% per annum (in thousands of rand at year 2006 prices).

Year	Investment cost	Recurrent cost	Benefits from reductions in health costs	Benefits from improved access	Net cash flows	Discount factor at 3% per annum	Present value at 3% per annum
0	−18				−18	1.00	−18.0
1	−18				−18	0.97	−17.5
2		−0.4	4.6	9.6	13.8	0.94	13.0
3		−0.4	4.6	9.6	13.8	0.92	12.6
4		−0.4	4.6	9.6	13.8	0.89	12.3
5		−0.4	4.6	9.6	13.8	0.86	11.9
6		−0.4	4.6	9.6	13.8	0.84	11.6
7		−0.4	4.6	9.6	13.8	0.81	11.2
8		−0.4	4.6	9.6	13.8	0.79	10.9
9		−0.4	4.6	9.6	13.8	0.77	10.6
10		−0.4	4.6	9.6	13.8	0.74	10.3
Total	−36	−3.6	41.4	86.4	88.2		68.8
Present value at 3% per annum	−35	−3	35	73	69		69

Table A1.3 The present values of costs and benefits of an imaginary project at a discount rate of 20% per annum (in thousands of rand at year 2006 prices)

Year	Investment cost	Recurrent cost	Benefits from reductions in health costs	Benefits from improved access	Net cash flows	Discount factor at 20% per annum	Present value at 20% per annum
0	−18				−18	1.00	−18.0
1	−18				−18	0.83	−15.0
2		−0.4	4.6	9.6	13.8	0.69	9.6
3		−0.4	4.6	9.6	13.8	0.58	8.06
4		−0.4	4.6	9.6	13.8	0.48	6.7
5		−0.4	4.6	9.6	13.8	0.40	5.5
6		−0.4	4.6	9.6	13.8	0.33	4.6
7		−0.4	4.6	9.6	13.8	0.28	3.9
8		−0.4	4.6	9.6	13.8	0.23	3.2
9		−0.4	4.6	9.6	13.8	0.19	2.7
10		−0.4	4.6	9.6	13.8	0.16	2.2
Total	−36	−3.6	41.4	86.4	88.2		13.4
Present value at 20% per annum	−33	−1	15	32	13		13

Table A1.4 The present values of costs and benefits of a water project at a discount rate of 30% per annum (in thousands of Rand at year 2006 prices).

Year	Investment cost	Recurrent cost	Benefits from reductions in health costs	Benefits from improved access	Net cash flow	Discount factor at 30% per annum	Present value at 30% per annum
0	−18				−18	1.00	−18.0
1	−18				−18	0.77	−13.8
2		−0.4	4.6	9.6	13.8	0.59	8.2
3		−0.4	4.6	9.6	13.8	0.46	6.3
4		−0.4	4.6	9.6	13.8	0.35	4.8
5		−0.4	4.6	9.6	13.8	0.27	3.7
6		−0.4	4.6	9.6	13.8	0.21	2.9
7		−0.4	4.6	9.6	13.8	0.16	2.2
8		−0.4	4.6	9.6	13.8	0.12	1.7
9		−0.4	4.6	9.6	13.8	0.09	1.3
10		−0.4	4.6	9.6	13.8	0.07	1.0
Total	−36	−3.6	41.4	86.4	88.2		0.0
Present values at 30% per annum	−32	−1	11	22	0		

the internal rate of return measure is that it gives a figure which people can compare with the interest rate that they get on their savings (or which they pay on loans). Clearly, a real rate of return of 30% per annum is very attractive.

THE THREE MEASURES BROUGHT TOGETHER

The imaginary project gives a benefit/cost ratio (when the investment costs are annualized at an interest rate of 3% per annum) of 3, so it would seem to be attractive. The net present value of the project is positive at a discount rate of 3% per annum and again positive at 20% per annum. The internal rate of return is estimated to be 30% per annum, again very attractive.

We have three measures to choose from. The measure with which most people can connect is the internal rate of return, since this can be compared with the interest rates which people receive on deposits or pay on loans.

However, the choice measure is less important than having an idea of its reliability. Any assessment should include *sensitivity analysis.* This consists of seeing how the measure changes when changes are made to the estimates of particular costs and benefits.

SENSITIVITY ANALYSIS

If we assume that, for this imaginary project, we want to value the savings in access time not at GDP per capita but at much less, say 30% of GDP per capita. Looking back at Table A1.2, we can see that the present value of the benefits from improved access is rand 73 000. If we take 30% of this, we have rand 22 000, and, therefore, the benefits are reduced by rand 51 000. Whereas the previous net present value was rand 69 000, it is now rand 18 000. The net present value is still positive at 3% per annum. However, the health benefits would also be reduced, as some of these are time savings, by somewhere between 0 and 70%. Even if they are reduced by 70%, they would still be reduced by "only" 70% of rand 35 000 or by rand 17 000. The net present value would still be slightly positive.

It is clear that by calculating the present values for each of the costs and benefits (as is done in Tables A1.2, A1.3 and A1.4), sensitivity analysis is facilitated.

REFERENCE

Hutton G, Haller L, Bartram J (2006). *Economic and health effects of increasing coverage of low-cost water and sanitation interventions*. Occasional Paper for the 2006 UNDP Human Development Report.

ANNEX 2

THE VALUATION PROBLEM

Given the importance of the benefits from reduced access time resulting from both water and sanitation improvements, both the quantity of time savings as well as their valuation need careful examination. This annex to Chapter 12 elaborates on the problem of valuation.

The first and obvious question to ask in relation to the valuation of benefits is; why not use market prices?

Many people in developing countries buy at least some of their water from water vendors. So can we not use the prices that they pay as a measure of the benefits of water? As we have seen in Chapter 3, many poor households spend a large proportion of their income on water. For example, the poorest 20% of households in El Salvador, Jamaica and Nicaragua spend more than 10% of their household income on water whereas a level of 3% is seen as an indicator of hardship in the United Kingdom (UNDP, 2006). However, the use of actual market prices (for example, as paid to water vendors) is likely to be misleading because in general only part of the water supply is purchased.

The next question might be to ask: if market prices do not exist, why not use the prices which people say they are willing to pay? This approach is known as willingness to pay (or contingent valuation) and is surprisingly widely used given its severe limitations. These are discussed by Hutton (2000). The major problem is that willingness to pay in prospect is not necessarily the same as willingness to pay in retrospect. Hutton (2000) cites a study by Boadu in 1992 which found that those households that had recently experienced a water-associated illness were willing to pay more for water than those households that had no recent history of water-related illness.

Because market prices do not exist for most of the water supply and because willingness-to-pay is unreliable, we have to adopt some other shadow price, that is a price which does not exist and yet is thought to represent the value to a whole society of a resource or commodity that is being provided.

One way of arriving at a shadow price is by approaching the problem through valuing time used for other uses rather than from the demand (or willingness–to–pay) side. This approach is the one most commonly used in the cost–benefit analysis of those services for which there are no market prices or for which market prices are thought to be an inadequate reflection of social values. It is the approach used by Hutton, Haller & Bartram (2006) for analysing water and sanitation services. They state that an "advantage of cost-benefit analysis over a purely financial analysis is that a proxy value of time can be used and applied irrespective of what individuals actually do with their time".

Hutton, Haller & Bartram (2006) value time saved at the GDP per capita of the country concerned, with the annual GDP being transformed into an hourly value. They give two reasons for this. One is equity, namely that "it is appropriate to assign to all adults the same economic value of time gained". The second is that people value their time at or close to the average hourly wage in their economy. Hutton, Haller & Bartram say: "….based on the above evidence and considerations, the GNP per capita (in US$) in the year 2005 is used as the average value of time in an economy".

The reasoning of Hutton, Haller & Bartram (2006) – 'irrespective of what individuals actually do with their time' – is likely to be heavily criticised by many cost–benefit analysts on the grounds that if the individuals are unemployed, the opportunity cost of their time could be considered to be zero or close to zero. The opportunity cost is given by the value of the next best use of their time. In other words, if there is little or no addition to production as a result of the time saving because the person is unemployed, then the value of the time is zero or close to zero. Some concession might be made to allow for the valuation of leisure time, but even then the value is likely to be much less than the average GDP per capita. Hutton & Haller (2004) conceded that their valuation of time (equal to the minimum wage) may overestimate the actual economic value, because of the presence of unemployment or underemployment.

As we have seen, one argument to counter this zero opportunity cost could be that it is equitable to assign to all people the same time value. However the equity argument is likely to get little sympathy from neo–liberal economists for two reasons. First (neo–liberal economists are likely to argue), in a market economy, individuals are paid according to the value of their production and therefore the grounds for treating all on the same basis are inequitable. A second objection is likely to be: why should resources be redistributed in the form of water and sanitation facilities aimed at the non-productive people? Some of the arguments for and against applying weights for income distribution are discussed in Box A2.1.

Box A2.1 Income distribution and cost–benefit analysis

A common approach in cost–benefit analysis is to value time according to its opportunity cost – that is, according to the production gained or lost. This will usually mean that time will be more highly valued for some workers than others. For the unemployed this may mean a valuation of time at close to zero. Similarly, it is common in cost–benefit analysis to value a life by the future production lost. Thus, the life of a highly-paid person is often valued more highly than the life of a lowly-paid person of the same age.

Some economists feel very uncomfortable about this and argue that it is desirable to handle income distribution in cost–benefit analysis by attaching compensatory weights to the incomes of people with a low income. This is discussed by Curry and Weiss (2000). They consider a weighting system in which project benefits going to low-income groups are more highly valued than project benefits accruing to the high-income groups. As Curry and Weiss (2000) observe although distribution weights are widely discussed in theory, in practice they are rarely applied. This is partly because of the additional data requirements and partly because of the subjective nature of the value of the income distribution weights.

Many project analysts argue against the whole weighting process on the grounds that, if a government is unwilling to redistribute income through the tax system, why should it be done through project selection? Possible answers to these objections include:

- it may be cheaper to redistribute income more selectively through projects than through taxes;
- if the central government is unable to redistribute income through taxes, then it may want to do so through projects.

As Little & Mirrlees (1974) put it: "Our belief is that most governments would be happy if some quantified allowance for inequality were made. But many might prefer that it was done in a concealed manner so that the weighting system did not become the subject of parliamentary or public debate. It may sometimes be politically expedient to do good by stealth".

Of course, the approach of Hutton, Haller & Bartram (2006), namely to value the time of all adults at the same rate, has the virtue of simplicity. Given that, as already pointed out, access to safe water is a human right in some national constitutions, it also has considerable moral validity. Such a concern with social justice puts economics in its appropriate place in advising public deliberations rather than determining decisions.

REFERENCES FOR THIS ANNEX

Curry S, Weiss J (2000). *Project analysis in developing countries*, London, MacMillan

Hutton G (2000). *Considerations in evaluating the cost-effectiveness of environmental health interventions.* Geneva, World Health Organization. (WHO/SDE/WSH/00.10)

Hutton G, Haller L (2004). *Evaluation of the costs and benefits of water and sanitation improvements at the global level.* Geneva, World Health Organization. (WHO/SDE/WSH/04.04)

Hutton G, Haller L, Bartram J (2006). *Economic and health effects of increasing coverage of low-cost water and sanitation interventions*. Geneva, World Health Organization. Occasional Paper for the 2006 UNDP Human Development Report.

Little IMD, Mirrlees J (1974). *Project appraisal and planning for developing countries*. New York, Heinemann

UNDP (2006). *Human Development Report; beyond scarcity: power, poverty and the global water crisis.* New York, United Nations Development Programme

Index

Withdrawn from University of Brighton Libraries